T0250020

Perry C. Crandall, PhD

Bramble Production: The Management and Marketing of Raspberries and Blackberries

*Pre-publication
REVIEWS,
COMMENTARIES,
EVALUATIONS . . .*

"**I**f you are involved in any aspect of bramble production, you need this book in your library. Perry Crandall has prepared a very thorough and practical guide to the history, biology, culture, and management of raspberries and blackberries. The book takes the reader through every aspect of bramble production in a clear and detailed presentation written in an enjoyable style. This book will quickly establish itself as one of the first sources to check for information, and it is destined to become a classic reference."

David G. Himelrick, PhD
*Associate Professor of Horticulture,
Auburn University*

More pre-publication
REVIEWS, COMMENTARIES, EVALUATIONS . . .

"**B**ramble Production is the first comprehensive treatise of raspberry and blackberry culture. It will serve as an essential reference for growers, consultants, and researchers alike."

William P. A. Scheer, MS
Area Extension Agent,
Washington State University

"**B**ramble Production is a long-overdue, definitive update on the cultivation and production of brambles illustrating practices currently in use in the Northwest. Dr. Crandall's credentials are outstanding, and he is highly respected in the industry. I predict that this book will become a standard work."

Ivan H. Matlock, BS
Owner/Operator of Far West
Nurseries, Inc.;
Partner, Matlock Partners

"**B**ramble Production does an admirable job of bridging the formidable gap between horticultural theory and bramble production in the real world. It is an excellent general reference. The new grower or student will appreciate step-by-step outlines of procedures, pictures, and tables of useful information. The experienced grower and professional horticulturist will look to this text to fill in 'knowledge gaps' for teaching and to obtain a 'second opinion' on a variety of topics.

Readers from many geographical areas of bramble production will glean useful information from these pages. Reading this book is like having an extended conversation over coffee with a friendly and incredibly competent horticulturist who has just been asked, 'Tell me everything you know about bramble culture.'"

J. Scott Cameron, PhD
Associate Scientist,
Department of Horticulture
and Landscape Architecture,
Research and Extension Unit,
Washington State University

"**B**ramble Production is excellent for anyone interested in gaining a working knowledge of the botany, growth, and production of brambles. It is easily understood by the novice, but contains the detail and practical information to make it a valuable reference manual for commercial producers, students, bramble researchers, and extension personnel."

John A. Lipe, PhD
Professor and Extension Horticulturist,
Fruits and Nuts,
Texas Agricultural Extension Service

"**D**r. Crandall's book embodies the results of over 30 years research experience with brambles in the Pacific Northwest. It is easy to read and is an excellent guide for new, inexperienced growers. The bramble production chapters are excellent and contain many useful tips for the beginner."

Adam Dale, BSc, PhD
Small Fruit Research Scientist,
Horticultural Research Institute
of Ontario

"**D**r. Crandall's book will prove a valuable reference for both commercial canefruit growers and students engaged in production horticulture. *Bramble Production* covers the gamut of commercial decision making, from site selection and soil preparation, to pest control and marketing, as well as sections on the physiology of growth and development, and the principles of pruning and training.

Most of the world's cane fruit crops are consumed in a processed form and could, potentially, be machine harvested. The never-ending quest for greater efficiency will force more producers, worldwide, toward machine harvesting. Dr. Crandall's discussion of machine harvesting and the postharvest handling and marketing of machine-harvested fruit will prove of particular value."

Graeme McGregor, MSc
Department of Agriculture, Victoria
Institute for Horticultural Development,
Australia

More pre-publication
REVIEWS, COMMENTARIES, EVALUATIONS . . .

"**B**ramble Production is written in such a reader-friendly way that interested growers can learn a great deal about the crops without being discouraged by scientific jargon. Overseas readers will also find much to interest them, since the author has ranged worldwide in his reviews, and his list of references is very extensive. This greatly increases the value of the book.

A vast amount of sound, commonsense information has been compiled in this guide to help growers make the right decisions on all aspects of crop management, production, and marketing. The grower can dip into the book for answers to particular queries or problems, as well as read up on complete subject areas. *Bramble Production* is a very worthwhile addition to the literature on cane fruits and should be of interest to a wide range of readers both in North America and abroad."

H. M. Lawson, MAgSc
Weed Scientist
and Raspberry Agronomist,
Scottish Crop Research Institute,
Dundee, Scotland

"**T**his volume provides complete up-to-date coverage on all aspects of the growing, management, and marketing of raspberries and blackberries. The author has done an admirable job of explaining the similarities and differences inherent in this group of plants. His knowledge of the subject, garnered over many years of research on these crops, is apparent.

Bramble Production will be of value to anyone with an interest in raspberries and blackberries. It will be of special value to persons growing these crops, whether on large commercial acreages or in a home garden. A handy reference for extension and research personnel."

James N. Moore, PhD
Distinguished Professor
of Horticulture,
University of Arkansas

Food Products Press
An Imprint of The Haworth Press, Inc.

NOTES FOR PROFESSIONAL LIBRARIANS AND LIBRARY USERS

This is an original book title published by Food Products Press, an imprint of The Haworth Press, Inc. Unless otherwise noted in specific chapters with attribution, materials in this book have not been previously published elsewhere in any format or language.

CONSERVATION AND PRESERVATION NOTES

All books published by The Haworth Press, Inc. and its imprints are printed on certified ph neutral, acid free book grade paper. This paper meets the minimum requirements of American National Standard for Information Sciences–Permanence of Paper for Printed Material, ANSI Z39.48-1984.

Bramble Production
The Management and Marketing of Raspberries and Blackberries

FOOD PRODUCTS PRESS

An Imprint of The Haworth Press, Inc.
Robert E. Gough, PhD, Senior Editor

New, Recent, and Forthcoming Titles:

The Highbush Blueberry and Its Management by Robert E. Gough

Glossary of Vital Terms for the Home Gardener
by Robert E. Gough

Seed Quality: Basic Mechanisms and Agricultural Implications
edited by Amarjit S. Basra

Statistical Methods for Food and Agriculture edited by Filmore
E. Bender, Larry W. Douglass, and Amihud Kramer

World Food and You by Nan Unklesbay

Introduction to the General Principles of Aquaculture
by Hans Ackefors, Jay V. Huner, and Mark Konikoff

Managing the Potato Production System by Bill B. Dean

Marketing Livestock and Meat by William Lesser

The World Apple Market by A. Desmond O'Rourke

*Understanding the Japanese Food and Agrimarket: A Multifaceted
Opportunity* edited by A. Desmond O'Rourke

Marketing Beef in Japan by William A. Kerr et al.

*Bramble Production: The Management and Marketing
of Raspberries and Blackberries* by Perry C. Crandall

Bramble Production
The Management and Marketing of Raspberries and Blackberries

Perry C. Crandall, PhD

Food Products Press
An Imprint of The Haworth Press, Inc.
New York • London • Norwood (Australia)

Published by

Food Products Press, an imprint of The Haworth Press, Inc., 10 Alice Street, Binghamton, NY 13904-1580

© 1995 by The Haworth Press, Inc. All rights reserved. No part of this work may be reproduced or utilized in any form or by any means, electronic or mechanical, including photocopying, microfilm and recording, or by any information storage and retrieval system, without permission in writing from the publisher.

Library of Congress Cataloging-in-Publication Data

Crandall, Perry C.
   Bramble production : the management and marketing of raspberries and blackberries / Perry C. Crandall.
      p. cm.
   Includes bibliographical references (p.) and index.
   ISBN: 978-1-56022-852-0
   1. Raspberries. 2. Blackberries. 3. Raspberries–Marketing. 4. Blackberries–Marketing.
I. Title.
SB386.R3C73 1994
634'.711–dc20   93-43912
                                                              CIP

CONTENTS

ABOUT THE AUTHOR

Perry C. Crandall, PhD, is Emeritus Professor of Horticulture at Washington State University, where he was on the faculty for twenty-nine years. He published nearly 100 scientific, research, extension, and grower-related papers on a wide range of horticultural crops. His main area of research was small fruits, and he is internationally recognized for his research with raspberries. Upon retirement in 1980, he was awarded the Distinguished Service to Agriculture Award by the Western Washington Horticultural Association. He is affectionately known among his colleagues as "Mr. Raspberry."

Since his retirement, Dr. Crandall has co-authored a chapter on raspberry management and served as a volunteer consultant for a blackberry grower in Guatemala under the auspices of the International Executive Services Corps. Currently, he is a member of the American Society for Horticultural Science, the American Pomological Society, and the International Society for Horticultural Science.

Both Perry and Margaret Taylor Crandall are natives of Iowa. They have been married for 53 years and have four grandchildren. Their home is in Vancouver, Washington.

Foreword

The commercial production of brambles is challenging and rewarding. Even though the cost of production is high, the potential profits are great. There are many different production practices available from which to choose, but no single combination of these is best for all situations. The success of the one that you choose depends on a thorough understanding of how the plants grow and respond to changes in their environment together with a knowledge of the different production practices available. This information will enable you to tailor a management system specially adapted to your own particular set of conditions and allow you to produce high yields with maximum profits.

The following chapters are meant to provide basic information on the botany, anatomy, physiology, and nutrition of brambles together with a discussion of the more popular management practices that are utilized by successful growers. Pertinent research results and other publications are referred to throughout the text. These offer inquisitive growers additional opportunities for even more intensive study.

In addition, it is important to keep up to date on changing trends in the industry. Maintain close contact with your agricultural advisors and attend grower meetings. There, you can visit with growers who have problems similar to your own and learn how they have been able to cope with them. You can listen to reports concerning the latest research on production practices, cultivars, market changes, and equipment. A good background in bramble physiology will enable you to evaluate these changes and predict how they might fit into your own operation. Continually keep in mind, however, that even though there are many cultural options available from which to choose, ultimately, the only right management combination is the one that works best for you.

Acknowledgments

I wish to give thanks to the following persons who have provided invaluable advice and support during the preparation of this manuscript.

To my faithful, patient wife of 54 years, Marna, who fed me, encouraged me, and gave me much appreciated advice.

To Dr. R. E. Gough, The Haworth Press Horticultural Editor, who spent many hours reading the manuscript and who gave me valuable assistance with format, content, and presentation.

To Ms. Lisa McGowan, Production Editor, who edited the entire manuscript and caught many typos, major and minor errors, and helped clarify ambiguous writing.

To Dr. C. A. Brun, Extension Horticulturist, who was an excellent source for illustrations and research material.

To Dr. Hugh A. Daubeny, Small Fruit Plant Breeder in British Columbia, who provided me with research information and unpublished manuscripts containing information that assisted greatly in the preparation of the chapter on plant selection.

To Dr. J. Scott Cameron, Washington State University Small Fruit Research Scientist, who gave me free run of his publication files and offered advice during the course of preparation.

To my granddaughter, Ann Carroll Pletcher, who prepared high quality, botanically-correct line drawings for the book.

To Jimmie D. Chamberlain, Research Technologist, who served as an invaluable assistant throughout my entire professional career in small fruit research.

To the many growers, scientists, field men, extension personnel, and scientists who, over the years, have given me ready access to their farms for observation and research and who have shared ideas and scientific information.

I have been blessed.

Chapter 1

Introduction and History of Use

The brambles include raspberries and blackberries and many hybrids (genetic combinations) of the two. The term "bramble" denotes thorniness, but there are cultivars (varieties) that are completely thornless. Wild species are found in every part of the world except the tropics. Jennings (1988) presented excellent descriptions of most commercially significant species.

Through the centuries, blackberries and raspberries have been valued for their vitamin content and outstanding flavor. They were gathered from plants growing wild in the woods and meadows to be eaten fresh, dried for storage, or canned. In many parts of the world today, both blackberries and raspberries still grow wild and are harvested for local use.

Red raspberries were grown by the ancient Greeks and used for medicinal purposes but it has been only since the latter part of the nineteenth century that they have been extensively grown commercially (Figure 1-1). The migration of large numbers of people from rural areas into the cities resulted in extensive urban sprawl and the clearing of more and more land for agricultural use. This isolated large groups of people from the delightful recreational opportunity to harvest and savor the wild fruit. These changes, along with the development of new, improved cultivars, better production and marketing practices, and market diversification, have greatly increased production and demand for the fruit. The result has been extensive increases in commercial acreage and the number of home garden plantings.

RED RASPBERRIES

Red raspberries are the most widely grown of the brambles. Currently there are large commercial acreages in Great Britain,

FIGURE 1-1. A healthy, well-kept, productive raspberry field. (Courtesy of Dr. C. A. Brun)

Europe, Canada, United States, Chile, New Zealand, and Australia. There are also many smaller commercial plantings and numerous home garden plots throughout the temperate zone ranging from Sweden and Finland in the north to Italy and California in the south.

The aboveground portion of brambles is biennial. In the spring, it develops new shoots from the basal buds of the canes or from buds on the roots, develops flower buds in the fall, and produces fruit the following spring. After fruiting, the canes die. During the first season, the shoots are called primocanes and in the second season they are called floricanes.

PRIMOCANE FRUITING RASPBERRIES

Some raspberries do not wait until the second season to produce fruit. They initiate flower buds on the terminal 1/3 to 1/2 portion of the primocanes during the first summer and fruit that fall. The lower

1/2 to 2/3 portion of the canes overwinters and produces a summer crop of fruit the following year. Cultivars with this growth habit are known variously as everbearing, fall fruiting, or primocane fruiting raspberries and have become quite popular as a means of extending the fresh fruit market into the high-priced, late fall market season. They can also be grown successfully in more northern regions where raspberry canes are often severely damaged by cold winter temperatures. Growers cut off these damaged canes at ground level during the winter. Although by doing this, they sacrifice the summer crop, they are still able to harvest a normal fall crop.

BLACK RASPBERRIES

Black raspberries are native to North America. They are not as winter hardy as red raspberries, are more susceptible to diseases, and are less productive. The fruit is firmer and has a very pleasing and distinctively different flavor. In the central and northeastern regions of the United States, their production is limited to relatively small acreages, but in the Pacific Northwest along the coast of Oregon they are grown on large farms and machine harvested for processing.

PURPLE AND YELLOW RASPBERRIES

Purple raspberries are hybrids that result from cross-pollination of red raspberries with black raspberries. They are intermediate in growth habit, have large fruit, and are juicier and more productive than black raspberries. Their production is almost entirely limited to the northeastern and central United States.

There are also yellow raspberry cultivars that, except for color, resemble red raspberries in appearance and flavor. Some commercial cultivars of these are available but they have never been very popular. They are primarily grown for specialty markets.

BLACKBERRIES

The early settlers of Europe and North America found wild blackberries growing in abundance (Darrow 1937). Although some were harvested for food, the majority, because of their thorniness and vigorous growth, were looked upon as a nuisance that interfered with land clearing and cultivation. It was only after plants with superior fruit were selected from the wild and propagated, that much interest in their domestic production developed. Even then, their thorny nature caused difficulty in handling and picking and slowed large-scale development (Figure 1-2).

About 1930, a thornless plant of the cut-leaf European blackberry was discovered (Darrow 1931). This selection was named 'Thornless Evergreen' and has been planted extensively in the coastal region of the northwestern United States where it thrives and produces huge crops. It does not produce well in the southern states because of a high chilling requirement (number of hours of temperature below 45°F [7°C] during the winter required to induce normal growth in the spring) and it is not winter hardy in the more northern states.

Plant breeders have, during recent years, developed a number of other thornless, high quality blackberry cultivars that are adapted to a wide range of climatic conditions (Moore 1980; Jennings, Daubeny, and Moore 1991). These, along with other new, improved cultivars, will result in further expansion of production both in the U.S. and abroad.

BLACKBERRY-RASPBERRY HYBRIDS

A perfect flower contains stamens (the male parts that produce the pollen) and pistils (the female parts that produce the seed). If the pollen from the stamens of one plant is transferred to the pistil of another plant it carries characteristics of the male parent (genetic material) and unites it with the genetic material in the pistil of the other plant. This produces a seed containing characteristics, part of which came from the male parent and part from the female parent. Plants produced from this seed are called hybrids and the process is the means whereby new cultivars are developed.

A blackberry-raspberry hybrid contains a mixture of characteristics from each of the two parents. 'Loganberry' is the first known example of such a hybrid. It was found in the garden of Judge Logan, an amateur horticulturist living in California. In 1881, he

FIGURE 1-2. Blackberry field in full bloom.

planted seeds from fruit of the 'Aughinbaugh' blackberry. Since this cultivar has only female (pistillate) flowers, the male pollen was assumed to have come from the flowers of the 'Red Antwerp' raspberries growing in the same garden. Subsequent controlled crosses involving blackberries and raspberries have yielded similar types of fruit (Jennings 1981).

'Loganberry' fruit is large, has an attractive conical shape, and is reddish-black in color. Even though its color and rather soft texture made it undesirable for freezing, it achieved immediate success. A thornless mutation of 'Loganberry' was discovered by B. E. and G. R. Bauer in 1929 (Brooks and Olmo 1944).

Over the years, 'Loganberry' has proved to be a very valuable tool for improving the native North American blackberry species (Jennings 1988). Although 'Loganberry' itself does not produce desirable hybrids when crossed with raspberry, other raspberry-blackberry crosses involving its progeny have been successful. Further improvements along these lines can be expected in the future with a major emphasis on the development of thornlessness.

UTILIZATION

Fresh Fruit Market

Bramble fruits are sold either fresh or processed. Large quantities of fresh fruit are sold right at the farm, either as fruit that is already picked or that which is available for customers to pick themselves. These methods of marketing are especially popular around large population centers and offer growers immediate cash at near retail market prices.

Large commercial farms and cooperatives have developed special precooling, packaging, and shipping practices that make it possible to send fresh fruit long distances. This fruit is carried by refrigerated trucks or by airplanes. The quantity of red raspberries handled in this way is increasing rapidly on the west coast of the United States and in South America, Great Britain, and New Zealand.

Processed Fruit Market

Many red raspberries are grown for processing in Scotland, Eastern Europe, New Zealand, and the Pacific Northwest region of North America. There are also large acreages of black raspberries and blackberries grown for processing in the state of Oregon. Most of the remaining blackberries are produced in California and other southern states.

Considerable quantities of fruit are mixed with sugar and packed in small, consumer-sized packages before freezing. Other fruit is individually quick frozen (IQF) by spreading it out on large trays in the freezer and then packaged. The majority is packed in large containers and frozen for later use in jams and jellies, pie fillings, yogurt, ice cream, juices and juice blends, wine, and other purposes. Only a small percentage of the production is canned and even less is used for dehydration.

HOME GARDEN PRODUCTION

Blackberries and raspberries are good crops for homeowners to grow in their garden. Given a well-drained soil and a temperate climate, they can be grown with little care and require minimal insect or disease control. The freshly-picked, home-grown fruit is unsurpassed for flavor and nutrition (Table 1-1). Most home gardeners only want enough fruit to top their morning cereal or ice cream, with some left over for State Fair blue-ribbon jelly or for canning and freezing.

TABLE 1-1. Composition and nutritive value of 100 grams of red raspberries or blackberries[1]

Nutrients	Raspberries	Blackberries
Water	87 g	86 g
Calories	49	52
Protein	0.9 g	0.7 g
Fat	0.6 g	0.4 g
Carbohydrates	12 g	13 g
Fiber	3 g	4 g
Ash	0.4 g	0.5 g
Minerals		
Calcium	22 mg	32 mg
Iron	0.6 mg	0.6 mg
Magnesium	18 mg	20 mg
Phosphorus	12 mg	21 mg
Potassium	152 mg	196 mg
Sodium	0 mg	0 mg
Zinc	0.5 mg	0.3 mg
Copper	0.1 mg	0.1 mg
Manganese	1.0 mg	1.3 mg
Vitamins		
Ascorbic Acid (Vitamin C)	25 mg	21 mg
Thiamin	0.03 mg	0.03 mg
Riboflavin	0.09 mg	0.04 mg
Niacin	0.9 mg	0.4 mg
Pantothenic Acid	0.2 mg	0.2 mg
Vitamin B_6	0.1 mg	0.06 mg
Vitamin A	130 IU[2]	165 IU[2]

1. From: *Composition of Foods, Fruits, and Fruit Juices—Raw, Processed, Prepared.* 1982. USDA Human Nutrition Service. Agr. Handbook No. 8-9.
   2. International Units.

Chapter 2

Classification of Raspberries and Blackberries

The main difference between raspberries and blackberries is in the way the fruit separates from the stem at the time of harvest. The raspberry loosens from the core, also known as the torus or receptacle, leaving it on the plant when harvested. This gives a fruit that is thimble-shaped with a hollow center (Figure 2-1). The blackberry, on the other hand, loosens at the base of the receptacle so the harvested fruit includes the receptacle as an integral part of it.

CLASSIFICATION

Plants and animals are classified into groups of individuals with similar natural characteristics using an international scientific classification system developed by the Swedish botanist Linné. This system separates the plant and animal kingdoms into Phylum, Class, Order, Family, Genus, Species, and sometimes the Variety in descending order from largest to smallest grouping. For example, the red raspberry is classified as follows: *Planta* (Kingdom), *Spermatophyta* (Phylum), *Angiospermae* (Class), *Rosales* (Order), *Rosaceae* (Family), *Rubus* (Genus), *idaeus* (Species).

All brambles belong to the Rose family, *Rosaceae*, along with apples, pears, strawberries, and many other plants with blossoms having five petals. In common practice, only the genus and species, and sometimes variety (subspecies), are used to specify a plant or an animal. Red raspberries are, therefore, usually referred to as *Rubus idaeus* L. The L. following the species designates the person who originally classified the plant, in this case, Linné.

FIGURE 2-1. Raspberries loosen from the receptacle when ripe forming a thimble-like fruit.

This binomial system of designation, i.e., genus and species, is used all over the world. If a Russian speaks of *Rubus idaeus* L., or in its abbreviated form as *R. idaeus*, a grower in China or New Zealand knows that he means the red raspberry.

The genus *Rubus* includes an enormous number of species besides raspberries and blackberries. Because there is such a diverse multitude of different ones, Focke (1910-1914) divided the genus into 11 subgenera, two of which include the brambles. He classified blackberries as subgenus *Eubatus* and raspberries as *Idaeobatus*. Some authors use this classification as being more inclusive but in this publication, both blackberries and raspberries are designated simply by genus and species with no mention of subgenus.

RED RASPBERRIES

All of the brambles belong to the genus *Rubus*. Red raspberries belong to the species *idaeus*, supposedly named after the region near Mount Ida in Asia Minor where they grew wild and were called "Ida" fruit by the ancient Greeks (Jennings 1988).

Most of the present day cultivars were developed from crosses of the subspecies (variety) *R. idaeus* var. *strigosus* (Michx.) and *R. idaeus* var *vulgatis* Arrhen. In recent years, *R. occidentalis* L., the black raspberry, has also been used to incorporate firmness, fruit rot resistance, and resistance to *Amphorophora idaei* Born., the aphid that transmits mosaic virus (Keep 1984).

Primocane fruiting raspberries are classified the same as other red raspberries. The tendency for fall fruiting already exists in some summer fruiting cultivars such as 'Lloyd George' and 'Ranere'. Plant breeders combined these cultivars with each other and with selections of *R. arcticus* and other wild species to obtain primocane fruiting cultivars with larger and earlier fruit and greater production (Keep 1984).

BLACK AND PURPLE RASPBERRIES

Black raspberries are found growing wild in North America. *Rubus occidentalis* L., the most common species, grows throughout the eastern United States ranging from New England south into the Carolinas. The species *R. leucodermis* Dougl. occurs along the Pacific coast from British Columbia south into California. Since most black raspberry cultivars have originated from the eastern species, all of them are classified as *R. occidentalis* L.

Purple raspberries are derived from crossing black raspberries with red raspberries. They are not usually given a specific Latin name, but are merely considered to be first generation hybrids. They are, however, sometimes listed separately as *Rubus neglectus* Peck.

BLACKBERRIES

Blackberries are often classified according to growth habit. They can be erect, semierect, or trailing, and either with or without thorns.

There is, also, a large and very important group consisting of hybrids of blackberry and raspberry. These, in turn, can be divided into the above classifications.

The erect and semierect species produce new plants from buds on the roots. The trailing types, often called dewberries, have few or no vegetative buds on the roots, instead, the tips of their primocanes form roots as they bend down and touch the soil. This is called tip layering and is characteristic of trailing blackberries and both black and purple raspberries.

There are numerous native species of blackberries spread over a wide range of climates and latitudes around the world. Over 350 different species have been identified (Bailey 1941-1945). Many of those used to develop modern day cultivars originated in either Europe or North America and have widely varied adaptation and characteristics. In addition, some outstanding cultivars involve blackberry by red raspberry crosses. Some of the more important species that have been utilized in the development of cultivars are listed in Table 2-1.

SPECIES NATIVE TO EUROPE AND THE MIDDLE EAST

R. procerus Muell. This species is native to Iran but has spread across southern Europe, England, New Zealand, and the western United States. Its thorny nature and excessive vigor have allowed it to escape from cultivation and become a very great nuisance in regions where it is adapted. The cultivated selections are often referred to as 'Himalaya' or 'Himalaya Giant'. The fruit are small and round, but the plants are quite disease resistant and very productive.

R. laciniatus Willd. *Rubus laciniatus* is named for its cut-leaf foliage and, like *R. procerus*, is thorny and vigorous. Because the plants retain their leaves through the winter, they are often called the 'Evergreen' blackberry. This species was introduced into western Washington and Oregon and is found growing wild throughout the region.

The 'Thornless Evergreen' cultivar, which is grown commercially in the Pacific Northwest, is a periclinal chimera of the wild species. In this type of mutation only the outer tissue of the stem mutates to the thornless condition. Laterals that arise from the stem

TABLE 2-1. World species of *Rubus* (raspberries and blackberries)[1]

SPECIES	NATIVE REGION	CHARACTERISTICS
RASPBERRIES		
R. idaeus var. vulgatus	Europe	Red raspberry, dark red, conic, few or no glandular hairs.
R. idaeus var. strigosus	North America	Red raspberry, light red, round, many glandular hairs.
R. occidentalis	North America	Black raspberry.
R. neglectus		Purple raspberry, a hybrid of red and black raspberries.
R. arcticus	Arctic regions of Europe and North America	Winter hardy, early, primocane fruiting.
BLACKBERRIES		
R. procerus	Europe	Small, round, thorny, vigorous, very productive.
R. laciniatus	Europe	Cut-leaf, trailing, evergreen, good quality, high yield.
R. thyrsiger	Europe	Good fruit quality.
R. nitidioides	Europe	Large, good flavor, early production.
R. rusticanus var. inermis	Europe	Major source of genetic thornlessness.
R. argutus	Eastern North America	Erect, source of genetic thornlessness.
R. allegheniensis	Eastern North America	Large, sweet, erect, thorny, winter hardy.
R. frondosus	Eastern North America	Large, sweet, erect, thorny, winter hardy.

TABLE 2-1 (continued)

R. baileyanus	Eastern North America	Good quality, trailing.
R. ursinus	Western North America	Trailing, crosses readily with raspberry, good flavor.
R. macropetalus	Western North America	Good flavor, trailing, crosses readily with red raspberry.
R. trivialis	Southeastern United States	Drought and heat resistant, low chilling requirement, trailing.

1. This table includes most of the more important species that have been used by plant breeders.

develop from the outer mutated tissue and are therefore thornless, whereas the suckers that grow from the roots originate from inner nonmutated tissue and are thorny. The gametes (the male and female cells that enter into the development of seeds) also come from the inner, nonmutated tissue. Therefore when *R. laciniatus* is used as a parent, it transmits the thorny characteristic to the seedlings.

R. thyrsiger Banning and Focke, and *R. nitidioides* Wats. The first is noted for its good fruit quality and the second for good flavor, large fruit, and early production.

R. rusticanus var. *inermis* E. Merc. This species is believed to have been involved in the development of 'Bedford Giant', the most popular blackberry in Great Britain. Crossed with *R. thyrsiger*, it produced 'Merton Thornless', a cultivar that is widely used in the breeding of thornless cultivars.

SPECIES NATIVE TO NORTH AMERICA

R. argutus Link, *R. allegheniensis* Porter, and *R. frondosus* Bigel. These species are all native to the eastern states. They are erect in growth habit and produce many primocanes from both roots and crowns. The fruit is large and sweet. They are used to produce upright, thorny cultivars suitable for colder climates and have been combined with 'Merton Thornless' to produce semierect, thornless cultivars.

R. baileyanus Britt. This trailing type is native to the eastern United States. It produces small clusters of good quality fruit and is believed to be one of the parents of both 'Lucretia' and 'Austin Thornless' dewberries. 'Lucretia' was a leading cultivar for many years and 'Austin Thornless' is often used by plant breeders to develop thornless cultivars.

R. ursinus Chamb. et Schlect and *R. macropetalus* Dougl. These west coast dewberries cross readily with blackberry species from other regions and with the red raspberry. As a result, many excellent cultivars have been produced from such combinations, the most important of which is the 'Loganberry'. This involved a cross between a blackberry with *R. ursinus* in its background, and a red raspberry. The fruit are reddish-black, large, long-conic, and have good flavor. The plants are vigorous and productive. 'Loganberry' has perfect flowers (flowers with both male and female parts in the same flower) and has played a very important role in transferring this trait to hybrids with the native western species which have the male and the female flowers on separate plants.

R. trivialis Michx. This southern dewberry species combines several characteristics that make it well suited for hot, subtropical climates. It has a low chilling requirement (the number of hours below 45°F [7°C] during the dormant season required to induce normal growth in the spring), will withstand hot summer temperatures, is drought tolerant, and resists spring frosts.

'Nessberry' was developed from a cross of *R. trivialis* with the red raspberry. Although 'Nessberry' was not successful as a commercial cultivar, it has been a valuable source of desirable traits for further breeding. It was used to produce the cultivar 'Brazos', an upright blackberry well suited to southern climates. 'Brazos' has been used extensively in breeding programs to produce other cultivars adapted to warm climates (Jennings 1988).

RASPBERRY-BLACKBERRY HYBRIDS

'Loganberry' and 'Nessberry' were two important early raspberry-blackberry hybrids. Other more outstanding hybrids have been developed since these were originated and the future holds great promise for even more.

Chapter 3

Growth and Development

Plants of the *Rubus* genus have many anatomical and physiological characteristics in common. All of them have flowers with five petals. The fruit is an aggregate of many tiny individual peach-like fruits surrounding the receptacle. Since the genus consists of a wide diversity of native species, it is not at all strange that there are other characteristics shared by most of the genus but for which there are exceptions.

Raspberries and blackberries have perennial roots and biennial tops. The roots continue to grow for the life of the planting, but new aboveground canes develop each year. Every year, the previous season's canes produce a crop of fruit and then die.

Sometimes this characteristic is modified by day length and temperature. Blackberries growing at a high altitude in Guatemala continue to produce flowers and fruit throughout the entire year. There, the canes do not die after fruiting but instead only the fruit laterals die back. The same cultivars growing farther north under different day-length and temperature conditions fruit seasonally and produce normal biennial canes.

Primocane fruiting cultivars also have a modified biennial habit. The primocanes produce fruit on the terminal 1/3 to 1/2 of the cane during the latter part of the first growing season. This terminal portion of the cane dies after fruiting, but the basal portion remains alive, overwinters, and produces fruit the second season, after which it also dies.

In the following discussion of growth and development, the emphasis is on red raspberries, since more information is available for them than for blackberries. Where differences in anatomy or the plants' response to environmental conditions occur, these differences are discussed.

MORPHOLOGY

Raspberries

Roots

Raspberry roots are concentrated in the upper levels of the soil (Colby 1936). Nearly 75% are present in the surface 1.5 ft (0.45 m) and they extend laterally in all directions. The remainder grow down as far as 6 ft (1.8 m) in well-drained soil. They consist of many fibrous roots and larger roots ranging up to 0.5 in (1.2 cm) in diameter.

The roots have many vegetative buds spaced at uneven intervals along them. During the first one or two years of the planting, many of these buds produce new shoots (Williams 1959a). These shoots develop roots and, while still young, they can be dug and used as a source of plants for establishing new plantings or to fill in where plants are missing. As the planting gets older, many of the root buds remain dormant and the number of new shoots that develop from them decreases. Those that grow between the hills and the rows compete for nutrients and water and must be controlled by either chemical or mechanical means. If left uncontrolled, they will fill in all of the area between rows, thus producing a solid jungle that is impossible to manage or harvest.

Black raspberry roots do not have these buds: new primocanes develop only from buds at and below ground level on the base of the plants (the crown area). Purple raspberries are intermediate in growth habit between reds and blacks and produce either none or a variable number of primocanes from root buds.

Primocane Growth and Development

Some root buds begin to grow in the fall (Hudson 1959; Williams 1959a). These grow to just above ground level, stop, form a rosette of leaves, and overwinter in this stage. In the spring, these along with shoots from other root buds and from basal crown buds elongate. During the first growing season, they are called primocanes. They grow rapidly, laying down axillary buds at the base of each leaf. There may be as many as three buds in a single leaf axil each at a different stage of development. The most advanced of these are called primary buds, the next advanced stage, secondary, and the least developed tertiary.

The distance between lateral buds varies, depending upon cultivar and vigor of the primocanes (Jennings and Dale 1982). It is greater on vigorous canes than on weak canes. Buds near the tip of the canes are close together since they are laid down late in the season under weather conditions that have slowed the rate of growth. The buds at the base of the cane just above and below ground level enlarge but do not develop further. These become primocane replacement buds for the following year (Figure 3-1).

During the summer of the first one or two years of the planting, the axillary buds on the primocanes of red raspberries often begin to grow and form laterals. In later years, they seldom form these laterals. However, if the growing points of the primocanes are

FIGURE 3-1. Crown area of plant showing replacement buds (a, b, c). (Illus. Ann Carroll Pletcher)

pinched off, as is usually done with black and purple raspberries and with erect blackberries, the primary axillary buds will break and develop laterals.

Primocane fruiting cultivars also produce laterals during the first summer. As soon as the terminal bud changes from a vegetative bud into a flower bud, terminal growth stops and laterals form on the upper portion of the primocane.

As temperatures drop in the fall and the days become shorter, the primocanes develop a rosette of leaves at the tip and stop growing. Black and purple raspberries continue to grow later into the fall than red raspberries. Their laterals bend over until the tips reach the soil at which time growth stops and they take root.

Change from Primocanes to Floricanes

As the growth of the primocanes slows in the late summer and fall, the lower temperature and shorter day length cause flower buds to be initiated in the terminal bud, changing it from a vegetative bud to a reproductive bud. This differentiation next occurs in the axillary buds at the base of the leaves beginning with those just below the terminal and continuing in order down the stem to ground level. Where cultivars form more than one bud in the leaf axil, the primary bud is the first to differentiate and remains the most advanced in development. Similar changes occur in the secondary and tertiary buds but the rate and degree of differentiation lags behind that in the primaries and may be totally lacking in some tertiaries.

Flower bud initiation and development takes place over an extended period of time in the fall, stops during midwinter and continues for a short period of time in early spring (Crandall and Chamberlain 1972; Dale and Daubeny 1987). By that time the individual flower buds have already developed microscopic sepals, petals, stamens, and pistils and are ready to expand into full bloom.

Primocane fruiting cultivars initiate the first flower primordia (the first stage of flower bud development) in the terminal bud of the primocanes during early summer (Vasilakakis, McCown and Dana 1979). This stops their terminal growth. However, the internodes below the tip continue to elongate for a short period of time but no new lateral buds are laid down and extension growth soon stops. At that time, the lateral buds on the upper portions of the primocanes begin to grow and produce lateral branches each of which develops terminal and lateral flower buds.

Floricane and Fruit Lateral Development in the Spring

The floricanes do not increase in length as growth begins in the spring. Instead, the lateral buds elongate and form fruit laterals bearing bracts, (immature, undeveloped leaves), leaves, and flowers. The laterals in the middle two-thirds of the cane are the most vigorous and fruitful (Locklin 1932). The terminal 10-12 in (25-30 cm) portion of the canes produces weak laterals with small, crumbly fruit. These tips are usually removed during dormant pruning.

The terminal flowers open first. Additional flowers develop in the axils of bracts immediately below the terminal flowers forming a loose, racemose inflorescence of three to five flower buds (a floral arrangement where the flowers are borne terminally and laterally on short flower stems along a central axis) (Reeve 1954; Williams 1959b) (Figure 3-2). Other racemose inflorescences occur in the axils of leaves below the terminal, each consisting of a terminal and subtending flower buds. The number that develop along each lateral and produce fruit is dependent on the vigor and nutrition of the individual laterals (Crandall, Chamberlain and Biderbost 1974).

Red raspberry flowers have long slender pedicels (flower stems) and the inflorescences occur in leaf axils extending well back from the terminal (Figure 3-3). The terminal buds bloom first, followed in turn by those farther down the lateral and even later on the lower laterals. This causes the fruit to ripen over an extended period of time and results in a long harvest season with a high yield potential.

The flowers of black and purple raspberries are borne on short, stiff, spiny pedicels concentrated near the tips of the laterals on closely spaced inflorescences. The fruit ripens over a short period of time. The harvest season is, therefore, short and the yield potential less than for red raspberries.

Flower Structure

Raspberry flowers (Figure 3-4) have five sepals, five petals, numerous stamens (pollen producing organs) and pistils (seed and

FIGURE 3-2. The arrangement of fruit on the laterals and a comparison of fruitfulness between two laterals. Some of the leaves have been removed to show the fruit better.

fruit producing organs). The stamens are arranged in two crowded whorls around the base of the receptacle, sometimes called the torus, (the enlarged terminal end of the pedicel which contains the flower parts) (Reeve 1954). There are about 150 pistils arranged spirally in several ranks above the stamens on the elongated receptacle. The base of each pistil enlarges to form an ovary containing two ovules (potential seeds), one of which develops. As the ovule grows, it is surrounded by a hard, flint-like seed coat. This, in turn, is covered with flesh and skin forming a miniature peach-like drupe. It requires the development of about 80-100 of these drupelets to form a normal fruit.

Fruit Structure

The fruit is an aggregate (results from a flower with multiple pistils which form tiny drupelets that are held together in a single

FIGURE 3-3. Racemose arrangement of raspberry blossom buds in the terminal inflorescence.

mass) consisting of 75-125 drupelets attached to a central receptacle (Figure 3-5). Each drupelet is an entire fruit. These drupelets are held tightly together by a dense entanglement of epidermal hairs (Reeve 1954; Robbins and Sjulin 1988). In some red raspberry cultivars and especially in black and purple raspberries there is some actual adherence of the flesh. The drupelets remain attached to the receptacle until they mature (Figure 3-6). At that time, they loosen from the receptacle and can be removed in the form of a thimble-like fruit, leaving the receptacle on the plant.

Fruit Growth and Physiology

Pollination

Fruit development begins with pollination. Nearly all bramble cultivars are self fruitful, that is, transfer of pollen from the stamen

FIGURE 3-4. Red raspberry flower cluster. The flowers consist of five sepals (a), five petals (b), numerous stamens (c) and 100-150 pistils (d). (Illus. Ann Carroll Pletcher)

to the pistil of the same flower or of another flower of the same cultivar results in a viable seed and formation of a drupelet. It requires 80+ drupelets to form a commercially acceptable fruit.

The pollen cannot be transferred by wind and only a very limited transfer takes place within the flower. Over 90% is done by flying insects, the majority of which are honeybees (Shanks 1969). Bees are strongly attracted to bramble blossoms so if they are present in adequate numbers and weather conditions are favorable for flight, they will do a good job of pollination.

Wild bees cannot be depended upon, especially during bad weather. Commercial growers, therefore, move hives of honeybees into the field during blossom time to insure a maximum crop. One or two hives per acre (2-5/ha) grouped into units of five or ten per location is generally recommended.

FIGURE 3-5. Closeup of red raspberry fruit.

Fruit Growth

The individual drupelets are anatomically and physiologically similar to peaches (Hill 1958). It requires about 30-35 days for red raspberry fruit and 35-45 for blackberries to mature after they are pollinated. The length of the harvest season is determined by the elapsed time from first to last bloom.

The rate of increase in size during the interval between pollination and harvest is not uniform. Growth is rapid during the first one-third, slows during the second third and speeds up again during the final third forming an S-shaped growth curve.

During the initial growth stage, the cells divide rapidly and form new ones. Practically all cells present in the mature fruit are formed during this period. The cells do not enlarge at this time and only about 10% of the final fruit weight accrues.

Cell division requires high levels of nitrogen and, if the supply is

FIGURE 3-6. Cluster of red raspberry fruit. Each fruit consists of 75 or more drupelets (a) surrounding the receptacle. The receptacle (b) remains on the plant when the fruit is harvested. (Illus. Ann Carroll Pletcher)

limited, the number of cells formed is less and fewer are available to enlarge in stage three.

The embryo within the seed develops and the seed coat hardens during the second stage. Size increase is limited to less than 5% of the total growth. Embryo development depends upon the successful pollination of the pistil. Without it, the embryo aborts and the drupelet fails to grow.

About 85% of the total size and weight of the fruit is determined during stage three. The cells laid down in the first stage enlarge, and sugars and other compounds accumulate causing this dramatic increase. Cell enlargement is highly dependent upon the supplies of carbohydrates (sugars) and water. Any limitation in the amounts available adversely affects the size of the fruit.

Since the size of the fruit of any particular cultivar is determined by the number and size of the individual drupelets; pollination, nitrogen, healthy foliage, and good light conditions together with water are the important factors to be considered in developing a management program.

Fruit Ripening

As the fruit ripens, it continues to increase in size and weight up to near harvest, the color changes from green to ripe-fruit color, flavor and sugar content increase, and the fruit softens and loosens from the receptacle. Ethylene production increases as the fruit ripens in both raspberries and blackberries (Jennings 1988; Walsh, Popenoe and Solomos 1983). The rate of CO_2 emission does not increase. Perkins-Veazie and Nonnecke (1992) found the same to be true with 'Heritage' raspberries and concluded that the fruit is non-climacteric. (In climacteric fruit, the CO_2 emission rate increases rapidly as the fruit matures.)

Blackberries

Roots

The root structure of blackberries is fibrous and relatively shallow like that of raspberries. Upright blackberries have vegetative buds on the roots, therefore, they produce primocanes from both roots and crowns. Trailing types have very few root buds and usually produce primocanes only from crown buds. Semierect cultivars are intermediate between the two and produce some primocanes from root buds but most of them come from the crown.

Primocane Growth and Development

The primocanes grow rapidly and with vigorous trailing cultivars, such as 'Thornless Evergreen', may reach a length of over 12 ft (3.6 m). A lateral bud is laid down at the base of each leaf. These do not normally develop into lateral branches until the second growing season, however, the terminal growth of erect types is pruned off during early summer. This causes them to immediately branch and form strong, fruitful laterals.

Most of the primocanes of newly planted erect blackberries grow horizontally on the ground during the first growing season. This often causes consternation among new growers who assume that this type of growth is what they can expect in future years. They can be assured, however, that after the first year the new shoots that develop will be stiff and erect as they should be.

Change from Primocanes to Floricanes

Primocanes continue to grow late into the fall when growth is stopped either by low temperatures or, in the case of trailing types, may be stopped by rooting of the terminals as they reach the ground.

The time when flower buds are initiated varies widely among blackberry cultivars, and depends upon cultivar, day-length, and temperature conditions. Robertson (1957) observed in Scotland that 'Himalaya Giant' began to form flowers in October while 'Ashton Cross' did not begin flower bud formation until mid-March. Waldo (1933) found that some late summer ripening types did not begin flower formation until early spring. Takeda and Wisniewski (1989) studied the time of flower initiation and rate of development for 'Black Satin' and 'Hull Thornless'. 'Black Satin' buds began initiation in the fall while those of 'Hull Thornless' did not form until spring. At higher elevations in Guatemala, the author found that the cultivar 'Brazos' formed flower buds and fruited the year around, probably as a result of low temperature and short day-length.

Floricane and Fruit Lateral Development in the Spring

Once buds begin to grow in the spring, the pattern of fruit lateral development closely follows that of the red raspberry. Early fruiting cultivars often have only five to ten fruits per lateral while some late fruiting types may produce as many as 50 or more.

Flower and Fruit Structure

Blackberry flowers consist of five sepals, five relatively large petals, and numerous stamens and pistils (Moore and Caldwell

1983). These are arranged on a fleshy, elongated receptacle (Figure 3-7). As the fruit matures, specialized cells form at the base of the receptacle blocking it off from the stem. When picked, the fruit separates from the stem at this abscission zone, thus the harvested fruit includes both a cluster of drupelets and the fleshy receptacle (Figure 3-8). This method of fruit separation is the principal characteristic differentiating blackberries from raspberries.

PHYSIOLOGY

Compensation

Raspberry plants have an unusually great capacity to compensate for reductions in yield components (Waister and Barritt 1980; Dale 1989). Each of the yield components has an effect on yield but for various reasons, their maximum potential is seldom realized. If one or more factors are adversely affected, other yield components tend to increase so that the total effect on yield is not nearly as great as would be expected. For instance, if the number of buds per cane is reduced, the remaining nodes become more fruitful. A reduction in cane number per unit area causes the remaining canes to produce longer laterals with more fruit per lateral and may increase fruit size (Crandall et al. 1974). Longer canes left at the time of dormant pruning result in more fruit per cane but fewer and smaller fruit per lateral. Other interactions are common. Almost all of them can be influenced by cultural practices that are under the control of the grower.

Yield Components

Total yield is determined by the number of fruit and size or weight of the fruit harvested. Fruit number is a result of the number of canes per lineal length of row, the spacing between rows, number of fruiting laterals per cane, blossoms per lateral, percentage of flowers that produce marketable fruit, the number of fruit actually picked during harvest, and the number discarded due to insect and disease damage. Fruit size is determined by the number of drupelets,

FIGURE 3-7. Cluster of blackberry flowers. Each flower consists of five sepals (a), five petals (b), numerous stamens (c) and 100-150+ pistils (d). (Illus. Ann Carroll Pletcher)

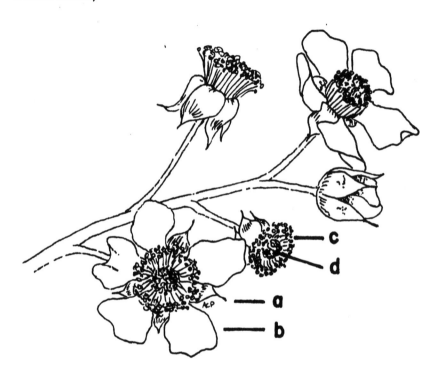

size of the receptacle in the case of blackberries, and the size of drupelets.

A grower can do much to influence yields through careful consideration of these yield components (Crandall 1980; Dale 1989). Management decisions and cultural practices can be changed or adjusted to improve production based on their potential effect on the various components of yield. Cultural practices that have a direct bearing are discussed in later chapters. Among the more important decisions and practices are site and cultivar selection, row spacing, soil management, irrigation, training and pruning systems, cover crops, pest control, mineral nutrition, and harvest methods.

FIGURE 3-8. Cluster of blackberry fruit. Each fruit consists of 75 or more drupe-lets (a) surrounding a succulent receptacle which separates from the base (b) and remains in the fruit when harvested. (Illus. Ann Carroll Pletcher)

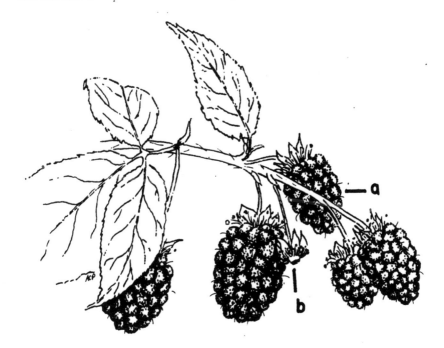

Chapter 4

Climatic Requirements

Weather often determines where brambles can be grown. Both summer and winter temperatures can be either too hot or too cold. A site can be too windy or it may lack sufficient air movement and the humidity can be too high or too low. In addition to these factors, day length interacts with temperature to influence production.

WINTER TEMPERATURES

State of Rest

The shortening days of fall and cool temperatures slow the rate of cane growth and cause vegetative buds to differentiate into flower buds. The canes stop growing and begin to acquire cold hardiness. This is a gradual process. Maximum hardiness is reached about one or two months after growth ceases. In northern temperate climates, this ranges from mid-November to mid-December. Biochemical changes occur within the plants that cause them to go into a state of rest, sometimes known as endodormancy. In this condition, they cannot resume normal growth even if temperatures become favorable. Raspberries and erect blackberries go into a state of rest earlier in the fall than do the trailing blackberries.

Chilling Requirement

Plants require an extended period of time at temperatures below 45°F (7°C) to overcome the rest period and allow the plants to resume normal growth. The total number of hours varies widely depending upon cultivar and species and is known as the chilling requirement for the particular cultivar. Failure to receive enough hours of low temperature to satisfy the chilling requirement results in poor lateral bud break in the spring and reduced yields. This limits cultivars with high chilling requirements to more temperate climates. 'Latham' red raspberry has a chilling requirement of about 1,400 hours whereas the newly developed cultivar 'Anita' (Rodriguez and Avitia 1989) requires only 250. Blackberry cultivars developed for the southern United States have low chilling requirements.

Raspberries are not well suited to southern or tropical climates because most cultivars have a relatively high chilling requirement (Hall and Brewer 1989). In warmer climates, the chilling requirement can sometimes be satisfied by growing the crop at higher elevations. Plant breeders have also made considerable progress in developing cultivars specifically adapted to warmer climates (Rodriguez and Avitia 1989).

Blackberries are better adapted to warm climates. Some of them have quite low chilling requirements and grow very well.

Dormancy

Before and after the rest period, growth can be halted at any time by unfavorable growing conditions. This physiological state is known as either dormancy or ectodormancy. When growing conditions once again become favorable, normal growth will resume. Varying degrees of dormancy can occur at any time throughout the growing season and is quite often the result of dry weather.

Winter Injury

In general, red raspberries are more hardy than black and purple cultivars, erect blackberries more hardy than trailing types and thorny cultivars more hardy than thornless types. Within each grouping there is considerable variation, much of which depends on the time of year and conditions within the plant when the cold weather occurs.

Unseasonal Fall Freeze

Cane growth slows and finally comes to a halt in the fall. As it does so, the plant slowly becomes more cold hardy until maximum hardiness is reached during the rest period. If unseasonal, below freezing temperatures occur during this acclimation period, the canes are likely to be injured. The lower the temperature and the earlier during this period that it occurs, the more likely that the plants will be severely damaged. This type of injury results in dieback or death of the canes. Cultivars that have late maturing canes, and those plantings that, because of poor management practices, continue to grow late into the fall are most subject to injury. Cultural operations designed to help prevent this type of winter injury are discussed in Chapters 9 and 12.

Injury During the Rest Period

Canes are most hardy during the rest period. The principal factor causing injury during this time is related to the genetic hardiness of the cultivar. Some red raspberries are quite hardy and can be grown successfully as far north as Finland and Sweden. Conversely, 'Brazos' blackberry is only hardy down to 0 to 5°F (–18 to –15°C) (Lipe 1986). Dana and Goulart (1989) set practical low temperature limits for red raspberries of –20°F (–29°C); purple raspberries, –10°F (–23°C); black raspberries, –5°F (–20°C); and blackberries, 0°F (–18°C).

Winter winds sweeping across exposed fields and plants that have been weakened by disease or other causes raise the temperature at which injury occurs. Injury during the rest period results in death or dieback of canes, crown damage, or death of the entire plant.

Late Winter Unseasonal Freezes

The canes gradually decrease in hardiness during the period from the maximum cold hardiness of the rest period until growth begins in the spring. Even a few days of temperature above 28°F (–2°C) decreases the hardiness of red raspberries during this time. Prolonged warm weather followed by a sudden cold spell increases the likelihood of injury (Figure 4-1). The degree of injury depends on how long after the rest period it occurs, the length of the warm period, and the extremity of the temperature difference.

Cultivars which are quite resistant to low midwinter temperatures, often suffer severe injury in regions where alternating warm and cold fluctuations occur during the spring. 'Latham' red raspberry is a good example. In Minnesota where winter temperatures are very cold, it thrives quite well. Farther south, in Iowa, where temperatures are less severe but fluctuate widely during late winter, it is often severely damaged.

Symptoms of this type of injury vary widely. They include dieback of canes, failure of lateral buds to develop, weak lateral growth, and normal initial growth of laterals followed by collapse at any time up until after fruit has formed. Examination of the lateral buds after this type of freeze reveals that the interior of the bud is

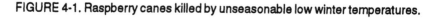

FIGURE 4-1. Raspberry canes killed by unseasonable low winter temperatures.

black or the bud itself may appear green and the conducting tissue at its base black (Doughty, Crandall and Shanks 1972).

Blossom-Time Freeze

A temperature of a few degrees below freezing at or near full bloom damages the blossoms and prevents fruit set. This type of injury is common in some locations with raspberries and early flowering blackberries. The nearer to full bloom that the freeze occurs, the more likely the damage. The tenderest parts of the flower are the pistils. With this type of damage, flowers develop black or brown centers one or two days after the low temperatures occur. Fruit does not develop, or in marginal situations, only a few drupelets form and the fruit is crumbly. Care in the selection of cultivars and the location of the planting will do much to prevent blossom-time injury.

SUMMER TEMPERATURES

Red raspberries do not thrive during hot, dry summers. A combination of heat, bright sun, and low humidity reduces fruit size and production and causes sunburn damage on exposed fruit. Such conditions exist in desert areas of both the north and south, though they tend to be most extreme in southern regions. Brambles growing in other areas with high summer temperatures and high humidity suffer also, but not usually as seriously. Some raspberry cultivars are more successful under such conditions than others. Blackberries are better adapted for these conditions but are not completely immune to injury. Careful scheduling of irrigation applications are helpful in reducing the scope of damage. Other suggestions for relief are included in Chapter 12 on fruit production.

WIND

Winter Wind

Strong, cold, dry winds desiccate the canes and increase the amount of damage caused by severe winter weather. These conditions often develop when subfreezing temperatures and windy conditions are accompanied by clear weather. The moisture in the air freezes out and reduces the humidity to very low levels. Strong winds during the fall and winter also cause cane breakage in exposed areas.

Summer Wind

Waister (1970), working in Scotland, found that yields were increased when the wind speed was reduced by windbreaks. The hot, dry, windy conditions often found in dry climates are even more likely to suppress yields. Careful scheduling of irrigation helps to alleviate the problem but yields are still adversely affected.

Day Length and Temperature

Long days (14 hr) and/or high temperatures favor vegetative growth while short days and cool temperatures cause flower bud initiation (Williams 1959b, 1960). A combination of short days and cool temperatures in the fall slows growth and causes the plants to go into a state of rest.

Chapter 5

Site Selection

SOILS

Brambles do best and can be grown most economically on fertile, deep, well-drained sandy loam or loam soils. Sandy soils are satisfactory, but they require frequent irrigations and special effort must be made to build up and maintain the organic matter. Be sure that there is a good supply of reasonably priced irrigation water readily available. Avoid heavy, poorly-drained clay soils and those locations that have a high water table during part of the year or that may be subject to flooding.

Brambles do not tolerate "wet feet." Plantings established on poorly drained soil are difficult to manage and, if the soil is saturated for extended periods of time, the roots suffocate from lack of oxygen and lose their ability to resist the invasion of root rot disease organisms, the plants grow poorly, and production is low. Many of the plants die.

Raspberry roots are especially sensitive to poor soil aeration. Blackberries are somewhat less sensitive, but all brambles require well-drained soils for best growth. As water fills the soil pores, it crowds out the air that is essential for respiration by the cells of the feeder roots. Respiration is the process whereby oxygen is combined with the sugars in the cells releasing energy and carbon dioxide. It is necessary for the active uptake of soil nutrients. Without adequate soil aeration, root growth is restricted and the plants wilt and grow poorly or die. They often develop abnormal nutrient deficiency symptoms.

Avoid soil that has been used for growing strawberries, peppers, tomatoes, potatoes, or eggplants during the past four to five years since they may be infected with *Verticillium* wilt. Soils with a history of *Phytophthora* root rot or crown gall, also, are not satisfactory. Soil fumigation kills these organisms but it is very expensive and cannot be relied upon to give long-term control.

TOPOGRAPHY

A location where the land is rolling or flat is best. Stay out of valley bottoms and away from southern slopes in cold climates. Cold air, being heavier than warm air, settles in the low areas and localized low temperatures are likely to kill blossoms. Southern slopes warm up first in the spring causing the blossoms to open early and making them vulnerable to late frosts. Steep slopes are subject to soil erosion, difficult to manage and are not suitable for machine harvest.

CLIMATIC CONSIDERATIONS

Air Movement

Avoid regions where strong hot or cold winds are common. Hot winds desiccate the fruit causing sunscald and greatly increase the amount of water used by the plants. Fruit size is reduced, plant growth is poor, and production low. Cold winter winds increase the danger of winter injury and often cause excessive cane breakage. Windbreaks can be used to reduce air movement but they should not be so thick that they cause dead air pockets within the field.

Some air movement is necessary. Lack of adequate ventilation causes high humidity levels within the foliage and results in increased cane disease and fruit rot development. Poor air circulation also causes frost pockets to develop in low areas resulting in blossom damage during cold spring nights.

MARKET

Locate commercial acreages as close to the market as possible. If the fruit is to be handpicked, be certain that there is a good supply of harvest labor available. On-the-farm marketing requires easy access to the farm together with a population density great enough to purchase the available fruit. Locate fresh market shipment plantings close to precooling facilities or build on-the-farm cooling.

Chapter 6

Plant Selection

Much of the success of a new planting depends on healthy planting stock. Brambles are subject to many viruses and cane and root diseases and special care must be used in choosing a source of plants. The use of certified plants is good insurance. Do not use plants dug from a neighbor's field. This is an invitation to trouble and can adversely affect the productivity and life of the planting.

Make the decision and order the plants well ahead of time, preferably the year before planting. This insures that you will receive the best quality plants and the cultivars of your choice.

CULTIVAR SELECTION

Select cultivars that are adapted to local climate and soil conditions. Consider the special requirements of the market for which the produce is intended. New growers can benefit by following the example of successful growers in the area who are producing fruit on similar soils and for the same type of market. Government agricultural advisors and processing company representatives are also good sources of information. Once the decision is made, order certified plants from a reliable nursery.

There is a large number of cultivars from which to select and new ones are being developed by plant breeders every year. Comprehensive listings of cultivars together with outstanding characteristics can be found in publications by Crandall and Daubeny (1990), Daubeny (1991), Moore and Skirvin (1990), and Jennings (1988). This wide variety of cultivars complicates the selection process, but it also provides an excellent opportunity to choose cultivars best suited for a specific market. Some are better for fresh market, some for processing, others can be used for both.

Characteristics of Cultivars

Over the years, cultivars have been selected from wild populations, from uncontrolled cross pollinated seedlings and, more recently, from controlled crosses designed to instill specific desirable qualities into them. The 'Thornless Evergreen' blackberry (Darrow 1931, 1937) is an outstanding example of a cultivar that was selected from the wild. It is still the most popular blackberry cultivar grown in the Pacific Northwest region of North America and one of the highest yielders (Jennings 1988). Many old cultivars were selected from open pollinated seedling populations as a result of the keen observation and efforts of amateur horticulturists who, when they observed an outstanding seedling, propagated it and made it available to others.

Most cultivars have a limited climatic range within which they are well adapted (Dale and Daubeny 1985). This means that, over the years, many cultivars have been named and introduced but only a few have been successful outside of a rather narrow range of conditions. Some of the factors that determine their range of adaptation are cold hardiness, chilling requirement, resistance to high temperatures, disease susceptibility, ability to grow well under a wide range of soil conditions, and insect resistance.

In addition to the above considerations, plant breeders have many other fruit and plant variables which they use in their selection of crosses to be made and in the evaluation of seedling populations. Some of the more significant fruit characteristics include size and shape, firmness, color, flavor, attractiveness, adaptability to different methods of processing, shelf-life after harvest, and ease with which the fruit can be removed from the plant. Plant growth characteristics that enter into the selection process are thornlessness, productivity, harvest season, resistance to insects and diseases, vigor, and growth habit. In general, each major growing region has developed one or two principal, especially well-adapted cultivars. Over the years, growers continue to plant new cultivars and plant breeders have continued to develop new cultivars that are superior to the old standards in one or more important characteristics. This has resulted in the complete or partial replacement of the old cultivars with ones better suited to the changing situation.

CULTIVARS FOR THE PACIFIC NORTHWEST REGION OF NORTH AMERICA

Red Raspberries

'Cuthbert' was the principle cultivar in this region from 1930 to 1940. It has been replaced by the larger-fruited, more productive 'Willamette' (Lawrence 1980) and 'Meeker', both of which can be machine harvested and are especially well adapted to the processing market. Promising new cultivars suitable for this market include 'Chilliwack' and 'Tulameen'.

The current emphasis of plant breeders in this region is on the development of cultivars for the fresh market, for firmness, and for insect and disease resistance. Promising cultivars from this emphasis include 'Chilliwack' and 'Tulameen', both mentioned above, plus 'Chilcotin', 'Skeena', 'Centennial', and 'Comox'.

Very few primocane fruiting raspberries are grown in this region. 'Heritage' is the principal cultivar but may soon be partially replaced by the earlier ripening 'Amity', 'Summit', or 'Autumn Bliss'.

Black raspberries are grown for processing in Oregon where 'Munger' is by far the most widely planted. In recent years, some 'Bristol' has been planted because of its larger size and better production.

Blackberries

The blackberry industry of the Pacific Northwest is concentrated in Oregon with some smaller plantings in mild climatic regions of Washington and British Columbia. 'Thornless Evergreen' and 'Marion' are the principal cultivars. Limited quantities of 'Chehalem', 'Loganberry', and 'Boysenberry' are also grown. Promising cultivars recently introduced include 'Silvan' and 'Waldo'.

CULTIVARS FOR CALIFORNIA

Raspberries

Red raspberries are grown in a very limited area around Watsonville in the coastal region of north central California. Nearly all of the fruit is harvested for fresh market and air shipped. The industry has successfully adapted the methods that proved so successful for marketing strawberries. Production is expanding rapidly. The principal cultivars are 'Heritage' and 'Willamette'. The latter cultivar produces both a summer and a fall crop under California conditions. 'Autumn Bliss', a British cultivar that produces higher yields and larger fruit, is being tested. New cultivars especially adapted to the region are being developed by local private breeding programs.

Blackberries

'Boysenberry' and 'Olallie' are grown for processing in the central valley of northern California. Both cultivars are productive, can be machine harvested, and have good processing quality. 'Olallie' ripens first followed by 'Boysenberry'. This is a stable industry and shows little evidence of change in the near future.

CULTIVARS FOR THE SOUTHERN UNITED STATES

Blackberries

Some scattered red raspberry plantings are grown in this region but they are not well adapted to the hot climate and warm winters. Erect types of blackberries bred specifically for Arkansas, Oklahoma, and Texas do well, but there is no major concentration of acreage. Most of the production is sold to the fresh market. Breeding programs are designed to develop low chill, thornless, erect cultivars adapted to machine harvest and resistant to double blossom, a fungus disease that seriously damages many of the commonly grown cultivars. Thornlessness is considered to be desirable for handpicking but not necessary for machine harvest.

'Shawnee' is the most widely grown cultivar in the southern part of the United States (Clark 1992). Other established cultivars in the region are 'Brazos', 'Cheyenne', and 'Cherokee'. These three are gradually being replaced by the newer cultivars 'Rosborough' and 'Choctaw' and the thornless types 'Navaho', 'Hull Thornless', and 'Arapaho'. 'Flordagard' and 'Gem' are grown in Florida and Georgia.

CULTIVARS FOR THE MID-WEST AND NORTHERN UNITED STATES, AND EASTERN CANADA

Bramble production in this region is almost entirely used for local fresh market. Most of the fruit is sold on farms, either as PYO (customers pick their own fruit) or already picked fruit. The areas in which the different types of brambles are grown is limited primarily by their ability to withstand cold winter temperatures. Red raspberries are the most cold hardy, followed by black and purple raspberries with blackberries as the least hardy.

Red Raspberries

'Latham' is the long-time standard cultivar for the region, primarily because of its winter hardiness and relatively wide adaptation (Lawrence 1980). It is still planted rather widely in the more northern states. 'Boyne', a new, hardy cultivar, is replacing it in the colder areas. Other old cultivars still being grown in the less cold regions are 'Milton', 'Taylor', and 'Newburgh'. Newer, promising cultivars include 'Titan', 'Nova', 'Hilton', 'Festival', 'Killarney', and 'Nordic'.

Primocane Fruiting Red Raspberries

'Heritage' is the most widely grown primocane-fruiting cultivar. It produces large, firm, good quality fruit on vigorous plants. By growing it for the fall crop only, it can be grown in the more northern climate. New, earlier ripening cultivars that show promise are 'Autumn Bliss', 'Ruby', 'Redwing', 'August Red', and 'Summit'.

Black Raspberries

Black raspberries are less winter-hardy than red raspberries and are less widely grown. 'Cumberland' and 'Bristol' are the two most popular cultivars. More recently developed cultivars include 'Blackhawk', 'Jewel', 'Logan', 'Morrison', and 'Allen'.

Purple Raspberries

Purple raspberries have never been very popular but with some of the larger, better flavored cultivars now available, they are increasing in popularity. 'Sodus' is the old standard but is being replaced by 'Marion', 'Brandywine', and 'Royalty'. The latter is considered to be the best purple of all.

Blackberries

Blackberry production in the midwest and northeastern United States consists of many small acreages located close to population centers where they can be sold locally. Cold winter temperatures confine them to the southern half of the region. Plant breeders continue to try to develop more cold hardy cultivars, but with only limited success.

'Darrow', 'Illini Hardy', and 'Hedrick' are erect, thorny types with the greatest cold hardiness. Less hardy, semierect, thornless cultivars include 'Smoothstem', 'Thornfree', and 'Black Satin'. New improved cultivars include 'Hull Thornless', 'Dirksen', and 'Chester'.

CULTIVARS FOR CENTRAL AND SOUTH AMERICA

The production of brambles in Central and South America is a recent development and consists primarily of red raspberries grown during the winter months for air shipment to the Northern Hemisphere. Most of the industry is concentrated in Chile.

'Meeker' and 'Heritage' are the principal cultivars grown. Some of the more recently developed North American and British raspberry and blackberry cultivars are being tested and show promise.

Among these are 'Autumn Bliss', 'Ruby', 'Chilcotin', and 'Chilli-wack'. 'Brazos', 'Cheyenne', and 'Rosborough' are grown in Guatemala and Costa Rica for air shipment. The recently introduced 'Navaho' and 'Arapaho' should prove popular there because of their thornlessness.

A small processing and fresh market blackberry industry has developed in southern Brazil using cultivars from the southern United States. 'Ebano', a thornless Brazilian cultivar, will likely become very popular in Brazil (Bassolos and Moore 1981).

CULTIVARS FOR NEW ZEALAND AND AUSTRALIA

Red Raspberries

The climate of New Zealand is well-suited to the production of bramble fruit, however, the industry is handicapped by the country's relatively small population and its isolation from the world market. Most of the fruit, therefore, is processed or sold on the local fresh market.

'Marcy' is the standard cultivar. It is very vigorous and productive under New Zealand conditions. 'Willamette' and 'Skeena' have partially replaced some of the acreage. There is, also, considerable interest in newer cultivars from the Pacific Northwest. 'Heritage' is popular and 'Autumn Bliss' is being tested as a partial replacement for it.

Southeastern Australia has a small, but growing, bramble industry. It is based on cultivars from the Pacific Northwest and Great Britain including 'Chilliwack', 'Skeena', "Nootka', 'Comox', and 'Glen Clova'. 'Heritage' is grown for a late season crop.

Blackberries

Blackberries are not grown widely in either New Zealand or Australia, although the acreage of 'Boysenberries' has increased considerably in New Zealand in recent years. Others which are grown are of the trailing type and include 'Marion' and 'Youngberry'. The new cultivar 'Silvan' is popular in south Australia (McGregor and Kroon 1984) and it is likely that growers in both countries will be interested in the thornless cultivar 'Waldo'.

CULTIVARS FOR GREAT BRITAIN

Red Raspberries

Just as is true in the United States, the predominant cultivars grown in Great Britain are those which were developed by plant breeders within the country. These come from the programs at the Scottish Crop Research Institute in Invergowrie and Horticulture Research International in East Malling.

'Glen Clova,' 'Malling Leo', and 'Glen Moy' are the principal cultivars. Others include 'Malling Jewell', 'Malling Admiral', 'Malling Delight', and 'Glen Prosen'. 'Autumn Bliss' is increasing in popularity as a primocane fruiting cultivar and 'Glencoe' is a promising new purple cultivar.

Blackberries

Blackberries are not extensively grown commercially in Great Britain because of the abundance of wild blackberries that grow throughout the countryside and the late ripening season of currently available cultivars. 'Bedford Giant' is grown to a limited extent in the south of England. The future of the industry depends on the development of early, thornless cultivars.

Blackberry-Raspberry Hybrids

New in the field, is a race of blackberry-raspberry hybrids of the 'Loganberry' type. First of these was the 'Tayberry'. It is being grown in Britain, North America, and Europe. Other more recent introductions include 'Tummelberry', 'Sunberry', and 'Fertodi Botermo'. The future for these hybrids looks good, especially if thornless cultivars can be developed.

CULTIVARS FOR WESTERN AND EASTERN EUROPE

Large acreages of red raspberries are grown in eastern and western Europe. Jennings, Daubeny, and Moore (1991) presented a good description of the rather complex cultivar situation in this region.

In addition to 'Willamette' and 'Heritage' from North America and 'Malling Exploit', 'Malling Promise', 'Lloyd George', and 'Autumn Bliss' from Great Britain, many cultivars have been introduced by breeders in western and eastern Europe. Of those from eastern Europe, 'Pavlovskaya', 'Unost', and 'Barnaulskaya' are popular in Russia; 'Bulgarian Rubin', in Bulgaria, the Czech Republic, Slovakia, and Poland; and 'Podgorina', in Yugoslavia. 'Veten' and 'Norna' are grown in the colder regions of Poland.

'Schoenemann' is the principal cultivar in Germany. It is also grown in France along with 'Jochem Roem' and some American and British cultivars. 'Veten' is the most popular cultivar in Norway and Sweden but it may soon be replaced by 'Balder', which is considered to be superior (Table 6-1).

TABLE 6-1. Raspberry and Blackberry Cultivars Around the World

---

RED RASPBERRIES

Algonquin From British Columbia. Winter hardy, very productive, upright, compact, numerous primocanes, aphid resistant, some resistance to spur blight and root rot. Fruit: Medium size, firm, bright red, nondarkening. Tends to cling to receptacle.

Amity From Oregon. Primocane fruiting, one to two weeks earlier than 'Heritage', vigorous, moderate production, some root rot resistance. Fruit: Medium size, very firm, medium dark red, good for processing and fresh markets.

Autumn Bliss From Great Britain. Primocane fruiting, erect, two to three weeks earlier and more productive than 'Heritage', aphid resistant, root rot resistant. Fruit: Large, slightly dark red, medium firm, mild flavor.

Balder From Norway. Very winter hardy, early. Fruit: Small, dark red, soft, good for processing. Hard to pick.

Boyne From Manitoba. Hardy, vigorous, sturdy, good yield. Fruit: Medium size, dark red, aromatic, only fair flavor.

Centennial From Washington. Productive, upright, vigorous, long laterals. Fruit: Large, conic, glossy medium red, medium firm, good for fresh market. Not suitable for machine harvest, lacks winter hardiness.

Chilcotin From British Columbia. Long harvest season, very good yield, upright growth. Fruit: Large, conic, bright glossy red, nondarkening, medium soft, good for fresh market. Not well adapted to machine harvest.

Chilliwack From British Columbia. High yield, upright, many primocanes, long strong laterals, hardy in British Columbia, aphid resistant. Fruit: Large, conic, glossy medium red, firm, very good flavor, easy to machine harvest, good for both fresh and processing markets.

Comox From British Columbia. Very high yield, hardy in British Columbia, vigorous, upright, numerous primocanes, suitable for machine harvest, aphid resistant. Fruit: Very large, conic, firm, bright medium red, good flavor, good for processing. Susceptible to root rot.

Festival From Ontario. Hardy, moderate yield, compact. Fruit: Medium size, firm; fair flavor.

Glen Clova From Scotland. Early, vigorous, very productive, numerous primocanes, long season. Fruit: Medium size, good flavor, firm, bright red, good for processing. Considered too vigorous without caneburning, susceptible to mildew.

Glen Lyon From Scotland. Moderate vigor, spineless, upright, excellent plant type, easy to manage. Fruit: Large, attractive bright red, firm.

Glen Moy From Scotland. Mid-early, very productive, moderate vigor, upright, short laterals, spine free, aphid resistant. Fruit: Large, bright red, firm, good flavor, good for both processing and fresh markets. Very susceptible to root rot.

Glen Prosen From Scotland. Mid to late season, upright, medium laterals, productive, aphid resistant, spine free. Fruit: Large, very firm, medium red, good flavor.

Heritage From New York. Primocane fruiting, vigorous, productive, upright, late season, widely adapted. Fruit: Medium size, attractive medium red, firm, fair flavor, good for freezing. Limited to warmer climates by its late ripening.

Hilton From New York. Vigorous, productive, mid-season, upright. Fruit: Large, conic, dark red, firm, good flavor.

Latham From Minnesota. Old standard, very hardy, vigorous, medium yield. Fruit: Small, round, light red, fair flavor.

Malling Admiral From England. Mid-season, vigorous, moderate yield, numerous primocanes. Fruit: Large, conic, medium red, firm, excellent flavor, suitable for fresh and processing markets.

Malling Delight From England. Mid-season, very vigorous, numerous primocanes, very productive, aphid resistant. Fruit: Very large, conic, orange-red, soft, fair flavor. Not suitable for processing, hard to pick.

Malling Leo From England. Late, very vigorous, upright, moderate yield, aphid resistant. Fruit: Large, round, firm, attractive, good flavor, suitable for fresh and processing markets.

Marcy From New Zealand. Old standard, vigorous, very productive. Fruit: Large, fair quality, suitable for machine harvest.

Meeker From Washington. Vigorous, productive, suitable for machine harvest. Fruit: Large, medium red, firm, good quality, suitable for fresh and processing markets.

Newburgh From New York. Old standard, productive, root rot resistant. Fruit: Large, firm, fair quality.

Nordic From Minnesota. Hardy, vigorous, productive, produces a very late fall crop. Fruit: Medium size and color, firm, good flavor, suitable for processing.

Ruby From New York. Primocane fruiting, slightly later than 'Heritage', moderate yield, primocanes do not branch. Fruit: Large, conic, bright medium red, good shelf life. Susceptible to root rot.

Schoenemann From Germany. Vigorous, productive, upright. Fruit: Large, good flavor, bright red, firm. Good for fresh market.

Skeena From British Columbia. Vigorous, productive, suitable for machine harvest. Fruit: Large, conic, medium red, suitable for both fresh and processing markets.

Summit From Oregon. Primocane fruiting, vigorous, productive, very early, root rot resistant, good for both fresh and processing markets. Fruit: Medium size, round, medium red, firm.

Titan From New York. Early, very productive, moderate vigor, not suited to machine harvest. Fruit: Very large, bright red, good flavor, soft. Susceptible to viruses and root rot.

Tulameen From British Columbia. Very productive, long late season, vigorous, suitable for machine harvest, good for both fresh and processing markets, aphid resistant. Fruit: Very large, conic, bright, medium red, firm, high quality.

Willamette From Oregon. Vigorous, numerous primocanes, widely adapted, disease and pest resistant, suitable for machine harvest, will produce a fall crop in warm climates. Fruit: Large, conic, dark red, firm, excellent for processing, good flavor. Susceptible to root rot.

BLACK RASPBERRIES

Allen From New York. Vigorous, medium yield, early, concentrated crop. Fruit: Large, firm, good quality. Susceptible to anthracnose.

Blackhawk From Iowa. Vigorous, medium yield, hardy, late, resistant to anthracnose. Fruit: Large, firm, good quality.

Bristol From New York. Very popular, productive, early. Fruit: Large, firm, glossy black, medium size, good quality. Susceptible to anthracnose.

Cumberland From Pennsylvania. Old cultivar, late, fair yield, midseason, susceptible to anthracnose, lacks hardiness. Fruit: Large, firm, good quality.

Jewel From New York. Early, productive, vigorous, resistant to anthracnose. Fruit: Large, firm, glossy, excellent flavor.

Munger From New York. Principal Oregon cultivar, early, suitable for machine harvest. Fruit: Medium size and yield, excellent for processing.

PURPLE RASPBERRIES

Brandywine From New York. Late, vigorous, productive. Fruit: Reddish-purple, large, firm, good for processing.

Glencoe From Scotland. Promising purple cultivar for Great Britain, vigorous, productive. Fruit: Large, attractive, good flavor.

Royalty From New York. Late, vigorous, high yield, immune to large raspberry aphid, suckers freely. Fruit: Very large, reddish-purple, sweet, very good flavor.

Sodus From Minnesota. Old standard, hardy, vigorous, upright, productive. Fruit: Large, firm, tart flavor.

ERECT BLACKBERRIES

Arapaho From Arkansas. Very erect, thornless, early, productive, suckers freely. Fruit: Good quality, firm, medium size, short conic, glossy black.

Brazos From Texas. Thorny, low chill, very early, vigorous, productive, spreading. Fruit: Very large, medium size, firm, glossy black, large seeds.

Cheyenne From Arkansas. Thorny, low chill, early, productive, resistant to orange rust. Fruit: Very large, firm, glossy black, excellent quality.

Darrow From New York. Thorny, vigorous, hardy, early, very erect. Fruit: Medium size, firm, good flavor.

Illini Hardy From Illinois. Thorny, late, vigorous, very cold hardy. Fruit: Medium size, glossy, good flavor.

Navaho From Arkansas. Thornless, late, productive, suckers poorly. Medium size, very firm, glossy black, very good flavor.

Rosborough From Texas. Thorny, early, very productive, low chill. Fruit: Very large, sweet, glossy black.

Shawnee From Arkansas. Thorny, very productive, late, immune to orange rust. Fruit: Very large, medium firm, good flavor.

SEMIERECT BLACKBERRIES

Black Satin From Illinois. Thornless, very late, vigorous, productive. Fruit: Large, firm, tart.

Chester From Illinois. Thornless, late, vigorous, cold hardy, very productive, resistant to cane blight. Fruit: Large, very firm, excellent flavor, resists hot weather.

Ebano From Brazil. Thornless, very late, low chill, productive. Fruit: Medium-large, firm, glossy, small seeds, excellent processing quality.

Hull Thornless From U.S. Dept. Agr. Thornless, very late, productive, very vigorous. Fruit: Large, firm, dull black, sweet.

TRAILING BLACKBERRIES

Bedford Giant From England. Thorny, very early, productive, very vigorous. Fruit: Very large, round, glossy, fair flavor.

Boysenberry From California. Thorny or thornless, late, very productive. Fruit: Large, soft, purple-black, excellent flavor for processing.

Kotata From Oregon. Thorny, early-season, vigorous, productive, suitable for machine harvest. Fruit: Large, glossy black, firm, excellent flavor, good for both fresh and processing markets.

Marion From Oregon. Thorny, early, productive. Fruit: Medium size, glossy black, excellent flavor.

Olallie From Oregon. Thorny, vigorous, midseason, very productive. Fruit: Large, firm, glossy black, good flavor.

Silvan From Australia. Thorny, very early, vigorous, productive, root rot resistant. Fruit: Large, glossy purple-black, good flavor, sweet, suitable for both fresh and processing markets.

Thornless Evergreen From Oregon. Thornless, late, vigorous, very productive, machine harvestable. Fruit: Large, glossy black, fair flavor.

Waldo From Oregon. Thornless, mid-early, high yield, machine harvestable. Fruit: Large, firm, glossy black, multipurpose, good shelf life.

BLACKBERRY-RASPBERRY HYBRIDS

Fertodi Botermo From Hungary. Semierect, thorny, productive. Fruit: Very large, good flavor.

Sunberry From England. Very thorny, semierect, productive. Fruit: Very large, long conic, dark purple, good flavor.

Tayberry From Scotland. Thorny, semierect, moderate vigor, productive. Fruit: Very large, chisel-shaped, excellent flavor, aromatic, dark red.

Tummelberry From Scotland. Semierect, thorny, productive. Fruit: Very large, round conic, medium dark red.

Chapter 7

Soil Preparation

PREPLANT PREPARATION

Site Preparation

Plan ahead. If at all possible, select the site at least one or two years ahead of time. This allows time to correct any problems that may be present before the crop is in the ground. It is much easier and cheaper to take care of them before planting.

Soil Fertility Test

The soil fertility test is basic to a good fertilizer program. Once the needs of the soil are known, it is possible to make the necessary soil fertility corrections required before planting and to develop plans for an ongoing fertilizer program.

A comprehensive soil test measures soil acidity, percentage of organic matter, levels of the major mineral elements phosphorus (P), potassium (K), calcium (Ca), and magnesium (Mg) as well as the minor elements (those nutrients essential to the growth of plants but required in very small quantities). These include boron (B), sulfur (S), iron (Fe), manganese (Mn), copper (Cu), zinc (Zn), and molybdenum (Mo).

Collect one or more soil samples from each of the different soil types in the field. The mineral content of samples taken from upland soils is different than that of those collected from lower, less well drained areas. Use a spade or soil sample tube to collect 15-20 samples of the 0-6 in (0-15 cm) layer of soil from scattered locations within the sample area. Place these in a clean pail, mix thoroughly and send a one- or two-cup sample to the laboratory for analysis. Your agricultural advisor can supply you with containers and complete instructions.

Soil Acidity

It is important to know the soil acidity level. Acidity is measured on a pH scale which ranges from pH 1 to pH 14. A pH below 7 is acid and one above 7 is alkaline. Raspberries and blackberries do best in a pH range between 5.5 and 6.5. The pH level can be adjusted to within this range by using agricultural lime to raise the pH or sulfur if it needs to be reduced (Table 7-1). These materials, if needed, are incorporated into the soil during the fall or winter of the year before planting.

Organic Matter

Brambles thrive in soils with organic matter levels above 3%. If the amount of organic matter is below this level, begin a program to

TABLE 7-1. Amount of agricultural lime or ground sulfur required to change the pH of soils

To Change	Agricultural Lime (lbs/A) or (kg/ha)		
from pH to:	Sand	Sandy Loam	Loam
4.5-6.5	2600 lb (2914 kg)	4200 lb (4700 kg)	5800 lb (6500 kg)
5.0-6.5	1800 (2000)	3400 (3800)	4600 (5200)
5.5-6.5	1200 (1300)	2600 (2900)	3400 (3800)
6.0-6.5	600 (670)	1400 (1600)	1800 (2000)
Ground Sulfur			
8.0-6.5	1200 (1300)	1400 (1570)	1500 (1700)
7.5-6.5	500 (560)	625 (700)	800 (900)
7.0-6.5	100 (110)	130 (145)	150 (170)

raise the level before planting and continue it during the life of the planting. Such a program may include any of the following sources of organic matter.

Animal Manure. Dairy cattle manure is an excellent source of organic matter. Incorporate it into the soil at a rate of 10-20 tons/A (22-44 MT/Ha) in the fall. Other types of manure are less desirable. Horse manure often has troublesome weed seeds. Chicken manure is low in organic matter and may cause plant damage because of high nitrate levels. Swine manure is low in organic matter.

Preplant Green Manure Cover Crops. A good cover crop produces a large amount of organic matter through both top and root growth, is easy to incorporate into the soil, does not produce seeds that carry over and act like weeds following incorporation, germinates and grows readily under a wide range of weather conditions, helps to smother germinating weed seeds, and does not promote soil borne diseases (Table 7-2).

One crop that most nearly satisfies all of these criteria is annual or perennial ryegrass. In situations where it can be grown, it should be the first choice of growers wishing to build up the soil organic matter. Other commonly used green manure crops fail to fulfill some of these requirements.

Alfalfa that has been growing in the field for at least two years and has a heavy enough stand to crowd out or prevent the growth of perennial weeds is a good alternative. It opens up heavy soils with its deep root system, breaks down rapidly once incorporated, and adds large quantities of nitrogen to the soil. It does not, however, contribute as much organic matter as ryegrass. Alfalfa or other broadleaved cover crops should not be used if dangerous populations of nematodes are present. Such plants actually favor the build-up of these pests.

Soil Test for Nematodes

Nematodes are microscopic, nonsegmented worms that feed on the roots of plants and organic matter. They invade the roots and can transmit ringspot virus from infected plants to noninfected plants. Other nematodes cause the formation of nodules on the roots that interfere with nutrient uptake and weaken the plants.

Collect soil samples for nematode analysis in the spring or early summer of the year before planting. This allows fall treatment for

TABLE 7-2. Green manure cover crops

Annual or Perennial Ryegrass

Seeding rate–30 lb/A (34 kg/ha). Germinates rapidly, grows well under a wide range of climatic conditions, produces large amounts of organic matter because of its extensive and fibrous root system with moderate top growth, is easy to incorporate and does not invade the bramble rows, can be planted from early spring through late summer.

Alfalfa

Seeding rate–14 lb/A (16 kg/ha). Requires one or two years to become fully established, grows best in well-drained more alkaline soils (pH 6.5 or above), liming may be necessary, roots grow deep and help loosen heavy soils, breaks down rapidly and releases large amounts of nitrogen upon decomposition, favors the multiplication of harmful nematodes, adds only moderate amounts of organic matter.

Wheat, Oats, and Barley

Seeding rate–100 lbs/A (112 kg/ha). Germinates and grows rapidly on most soils, will continue to grow through the winter in milder climates, needs to be incorporated in early spring, best used postharvest to slow growth and prevent erosion.

Sweet Clover

Seeding rate–12 lb/A (14 kg/ha). Biennial, requires soil pH of 6.5 or above, large roots that penetrate deep into the soil, heavy top growth which decomposes rapidly when incorporated, good source of nitrogen.

Field Brome

Seeding rate–20 lb/A (22 kg/ha). Winter annual, germinates and grows rapidly, heavy fibrous root system and strong spring growth, will reestablish in late summer if allowed to go to seed, ranks with ryegrass in amount of organic matter produced, more difficult to incorporate because of vigorous top growth.

Winter Rye

Seeding rate–100 lb/A (112 kg/ha). Commonly seeded during late summer, germinates and grows rapidly, overwinters in cold climates and makes vigorous spring growth, difficult to incorporate if allowed to grow tall, popular in more northern climates.

Cowpeas

Seeding rate–100 lb/A (112 kg/ha). Requires warm weather for best growth, produces large amount of organic matter in short period of time, easy to incorporate and breaks down rapidly.

Lezpedeza

Seeding rate–20 lb/A (22 kg/ha). Legume that is adapted to southern climates, heavy root and top growth, easy to incorporate and breaks down readily.

Hairy Vetch

Seeding rate–40 lb/A (45 kg/ha). Legume that is adapted to northern climates, often combined with winter rye to obtain more organic matter, the combination can be difficult to incorporate if allowed to grow late into the spring, provides a good supply of nitrogen, many hard seeds fail to germinate and carry over to act like weeds one or two years later.

---

their control if dangerous levels are found. Since nematode populations are not uniformly distributed but tend to be concentrated in "target areas" scattered throughout the field, special care must be taken to obtain a representative soil sample. A good sample consists of many small samples collected from the 0-6 in (0-15 cm) root zone of suspected areas of the field mixed together and sub-sampled for analysis. The experts in the laboratory will determine whether dangerous levels of nematodes are present and your farm advisor can make recommendations for their control.

Nematode control measures are expensive but failure to control them before establishment will jeopardize the success of the planting.

Tomato ringspot virus is transmitted from one plant to another by the dagger nematode *Xiphenema americanum*. Black raspberries are especially susceptible to this virus along with some red raspberry cultivars.

Apply the nematocide during late summer or early fall. Soil moisture levels in the spring are too high for good distribution of the fumigant throughout the soil and temperatures below 50°F (10°C) are too low for the nematocide to break down before time to plant.

Practice clean cultivation, either alone or in combination with chemical weed control during the summer before fumigation to control weeds and break down raw organic matter. Apply the fumigant to loose, well cultivated soil that has moderate levels of moisture and a temperature of between 55 and 75°F (13-18°C).

Soil Preparation

Perennial Weed Control

Eliminate perennial weeds during the summer before planting. Common, difficult to control weeds such as quackgrass (couchgrass, *Elymus repens* (L.) Gould), Canada thistle (*Cirsium arvense* Scop.), horsetail rush (*Equisetum arvense* L.), and Bermuda grass (*Cynodon dactylon* (L.) Pers.) are very difficult to control after the crop is planted. Use a combination of cultivation and chemical weed control for best results. Sometimes the herbicide can be successfully combined with an herbicide-resistant cultivated crop such as corn. Beware of herbicides that may remain in the soil over winter and damage the newly planted brambles.

Fall Plowing

Plow under the green manure cover crop in the fall along with necessary soil amendments such as manure, lime, or sulfur. Fall plowing helps the soil warm up and dry out sooner in the spring and allows earlier planting, which results in better establishment and growth during the first year.

In the spring as soon as the soil dries out enough to work properly, disc and harrow the field into seedbed condition. Soil that has the right amount of moisture crumbles when it is worked. Soil that is too wet forms clods and makes planting difficult.

Chapter 8

Planting

FIELD LAYOUT

Most commercial plantings are laid out in rows for ease of management and maximum production. Occasionally small commercial fields and home gardens are arranged in a square system where the plants are spaced an equidistance apart. These systems allow for easy control of weeds and removal of excess primocanes by cultivation (Figure 8-1).

A north to south direction of rows helps prevent sunburned fruit on the south sides of rows during hot summer days and promotes uniform fruit production on both sides of the rows. However, since long rows are desirable in commercial fields, the shape of the field usually determines the row direction. If rows are planted across the slope to prevent soil erosion, cultivation causes the rows to form terraces and makes it more difficult to harvest the crop, especially where the fruit is machine harvested. Noncultivation or minimum-cultivation soil management practices can be used in such a situation to reduce erosion and terracing. These techniques are discussed later in the section on soil management practices.

Row and Plant Spacing

The desirable spacing for plants between and within rows varies. Table 8-1 lists the maximum and minimum ranges. The space between rows can vary from a minimum of about 6 ft (1.8 m) to 12 ft (3.6 m). Since closer spaced rows produce higher yields per acre (Waister, Wright and Cormack 1980), space them as close together as convenient and practical. Wide spacing is sometimes necessary be-

FIGURE 8-1. A well-established, first-year raspberry planting during mid-summer. (Courtesy of Dr. C. A. Brun)

cause of machinery width. Space rows at least 2 to 3 ft (0.6-0.9 m) wider than the outside width of the machinery to be used.

Some cultivars are naturally more vigorous than others and often the same cultivar will grow much larger under more favorable climatic conditions. A cultivar trained on a crossarm or v-shaped trellis requires more space than on an upright trellis. Blackberries are usually spaced wider than red raspberries because they tend to be more vigorous. Black and purple raspberries are more spreading in growth habit and must also be spaced wider.

British growers often use 5-6 ft (1.5-1.8 m) row spacing for red raspberries (Turner and Muir 1985). They use specially built cultivation equipment, an upright narrow trellis system, favor less vigorous cultivars, retain a low population of primocanes per unit of row length, maintain the plants in hills or stools and keep fertilizer levels

TABLE 8-1. Recommended row and plant spacing

Crop	Space Between Rows	Space Within Rows
Red raspberries	6-10 ft (1.8-3.0 m)	2-3 ft (0.6-0.9 m)
Black raspberries	8-10 ft (2.4-3.0 m)	3-4 ft (0.9-1.2 m)
Purple raspberries	8-10 ft (2.4-3.0 m)	3-5 ft (0.9-1.5 m)
Erect blackberries	10-12 ft (3.0-3.6 m)	2-4 ft (0.6-1.2 m)[1]
Trailing blackberries	10-12 ft (3.0-3.6 m)	6-8 ft (1.8-2.4 m)

1. Blackberry root cuttings are often placed 2 ft (0.6 m) apart to fill in the rows more quickly.

relatively low. Even so, some of their more vigorous cultivars require wider spacing. Many growers are changing to 8 ft (2.4 m) or wider spacing to facilitate standard tractors and machine harvesters.

Raspberries grow very vigorously in the Pacific Northwest region of North America and in New Zealand. Primocanes are often 10 ft (3 m) long and growers usually leave ten or more canes per hill when pruning. They space the rows 8 to 10 ft (2.4-3.0 m) apart with rows to be machine harvested a minimum of 10 ft (3 m). Red raspberries grow much less vigorously in the eastern United States and Canada and, therefore, are often planted in rows 6 to 8 ft. (1.8-2.4 m) apart. Even there, wider spacing is sometimes used because of the width of available cultivation and spray equipment.

The space between plants in the row varies, depending on species, growth habit, and training method. For plantings to be allowed to develop into hedgerows, the plants are usually spaced 2 ft (0.6 m) apart to hasten the process. In fields to be maintained in hills or stools, they are spaced farther apart, usually 3 ft (0.9 m). Space black and purple raspberries 3-5 ft (0.9-1.5 m) apart because of their more spreading growth habit.

Erect blackberries are spaced 2-4 ft (0.6-1.2 m) in the row and the primocanes are allowed to fill in the space between plants. Use the closer plant spacing when root cuttings are used to establish the planting.

Trailing blackberries produce very long canes and are spaced 6-8 ft (1.8-2.4 m) apart. All trailing blackberries are maintained in hills and given enough space between plants for the canes to be trained on the trellises without overlapping.

Alleyways for Access to the Field

The maximum practical length of uninterrupted row for hand-picking is about 300 ft (91 m). This requires the establishment of 16-20 ft (4.9-6.1 m) wide breaks in the rows at this interval and similar width headlands at the ends of the rows. These alleys provide space for weighing and fruit accumulation and for access to remove the fruit from the field. For machine harvest, increase the distance between alleys to 1,000 ft (300 m) or more depending on the capacity of the machine to carry filled flats of fruit and the expected yield. Long rows and greater distance between stops to off-load the machine improve efficiency of machine harvest. Increase the width of the headlands at the ends of rows to 26 ft (8 m) to allow adequate turn around space.

ESTABLISHMENT

Marking Out the Field

Broadcast phosphorus and potassium fertilizer onto the soil in amounts specified by the soil test after the soil has dried out enough to break up readily. Then work it into seedbed condition.

Stake out the ends of the rows and the alleyways first. Then, if the field is to be handplanted and mechanically cultivated for weed control, mark each row with a deep furrow and handplant the plants in the bottom of the furrow at the proper spacing. As the plants grow, cultivate the soil to fill in around them and smother newly emerged weeds. Some growers mark each row and dig individual holes for the plants. Others have adapted mechanical transplanting machines which require that only the ends of the rows and the alleyways be marked out before planting.

Planting

With Dormant Plants

Order plants from the nursery well ahead of time to insure adequate quantities of the desired cultivars (Table 8-2). As the time for planting nears, notify the nursery of the desired time for delivery.

It is important that the roots not be allowed to dry out before planting. If the plants arrive too early, those that are packaged in

TABLE 8-2. Number of plants needed for various spacings[1]

Space Between Rows	Space Within Rows	Plants/A	Plants/Ha
6 ft (1.8m)	2 ft (0.6m)	3630	9259
6 ft (1.8m)	3 ft (0.9m)	2420	6173
8 ft (2.4m)	2 ft (0.6m)	2722	6944
8 ft (2.4m)	3 ft (0.9m)	1815	4630
8 ft (2.4m)	4 ft (1.2m)	1361	3472
8 ft (2.4m)	5 ft (1.5m)	1089	2778
10 ft (3.0m)	2 ft (0.6m)	2178	5555
10 ft (3.0m)	4 ft (1.2m)	1089	2778
10 ft (3.0m)	6 ft (1.8m)	726	1852
10 ft (3.0m)	8 ft (2.4m)	544	1389
12 ft (3.6m)	6 ft (1.8m)	605	1543
12 ft (3.6m)	8 ft (2.4m)	454	1157

1. The number of plants needed for other planting distances can be calculated by using the following formulas:

$$\frac{43560}{\text{ft between rows} \times \text{ft within rows}} = \text{plants per acre}$$

$$\frac{10000}{\text{m between rows} \times \text{m within rows}} = \text{plants per hectare}$$

polyethylene-lined boxes can be held in cold storage at about 35°F (8°C) for several weeks. Other plants should be heeled in. Dig a shallow trench in a sheltered location in well-drained soil. Open up the bundles of plants and space them in a single layer along the side of the trench. Cover the roots with moist soil well firmed down to keep them from drying out. Plants can be held in this manner for several weeks after which, depending on the level of dormancy and air temperature, they will begin to grow and must be handled very carefully to prevent loss of buds.

Take only the number of plants that can be planted in a half day or less to the field at one time. Cover the roots with wet burlap or immerse them in a container of water to keep them fresh.

Dig the hole large enough to allow the roots to be spread out. Cover the plants with soil to a depth of 2-3 in (5-8 cm). It is very important to firm the soil around the roots. During hot weather and in dry soil, apply enough water after planting to thoroughly wet the soil. Growers with large fields who use transplanting machines often follow immediately after planting with a good sprinkler irrigation to help settle the soil. They also have one or two persons follow along after the machine to fill in missing plants and reset those that are improperly covered.

After planting, cut off the "handle" (the piece of stem extending above the soil on tip-layered plants) close to the ground. This forces new growth from the crown or root buds and eliminates any cane diseases that are on the stub.

With Root Cuttings

Red Raspberries. Both red raspberries (Torre and Barritt 1979) and erect blackberries (Moore et al. 1978) can be grown from root cuttings. Though this method of establishing raspberries is not common, it is sometimes employed when certified stocks are in short supply.

Raspberry cuttings arrive from the nursery as a mass of roots of miscellaneous diameters and lengths and must not be allowed to dry out. Use 1 1/2 to 2 oz (44-60 g) of roots per 3 ft (0.9 m) of row. Either space them uniformly in a trench or at 3 ft (0.9 m) intervals in individual holes and immediately cover with soil to a depth of 1-2 in (2.5-5.0 cm). Apply water soon after planting to keep the soil moist until growth begins.

Erect Blackberries. Dig the roots early in the spring and either bury them in well-drained soil or hold them in plastic bags in cold storage at 32-35°F (0-2°C) until needed. Use only those roots that range from 1/4 to 3/8 in (5-9 mm) in diameter. Cut them into lengths of 3-4 in (7.5-10 cm) and place a single cutting horizontally at each plant location. Cover them with soil to a depth of about 3 in (7.5 cm). Water applied at the time of or soon after planting helps to settle the soil around the roots and gets the plants off to a rapid start.

Using Green Plants

Occasionally it is necessary to transplant plants which have developed green leaves and are in various stages of growth. Most often these are used to replace missing plants in an already established field. They may consist of root suckers, tip layers, or plants that have been grown from root cuttings in nursery rows or individually in plastic bags of soil. Take only the number of plants that can be transplanted within an hour or two to the field. Keep them moist and in the shade until needed. Replant them at the same depth that they were growing before and water in well.

Using Tissue Culture Plants

Tissue culture plants arrive growing in compartmentalized trays with a soil root ball. Water them as necessary and keep them in a sheltered location out of direct sunlight until needed. They can be planted by hand or with a transplanting machine. Cover the soil ball with about 1 in (2.5 cm) of soil and water it in. These plants begin growth soon after planting and grow rapidly during the first season and often out-yield fields propagated by other methods.

Replacement of Missing Plants

No matter how carefully the field is planted, there are bound to be some missing plants. It is important that these be replaced before the beginning of the second growing season to insure a full stand of plants since it is very difficult to fill in missing plants in older fields. Dig plants for this purpose from around the base of established healthy red raspberries and erect blackberries or layer enough tips of black and purple raspberries or trailing blackberries.

Chapter 9

Soil Management

The objective of a soil management system is to produce the maximum yield of high quality fruit with a minimum investment in labor and other costs. Soil type, weather, cultivars, labor availability, market and other variables all enter into the choice of which soil management program to select. No one system is best for all conditions and often the choice depends upon the desires and special needs of the grower. A grower who produces fruit for pick-your-own sale on the farm has different needs than a large grower who machine harvests his crop for the processing market. The ultimate soil management program involves many decisions from among the range of possibilities that are available. The one chosen need not, and probably should not, be final. As you gain experience, watch for ways to improve on it from year to year. New technologies often develop that can be incorporated. You or other growers often come up with new innovations. Progressive growers are always on the lookout for new and better ideas. It is important to keep up-to-date on the latest developments by studying research reports, extension publications and by attendance at pertinent grower conferences. One good idea picked up at a meeting or in reading can pay many times over for the time and money expended.

WEED CONTROL

The importance of eliminating troublesome perennial weeds prior to establishment of the planting was emphasized in Chapter 7. It is easier and more economical to kill them before establishing the planting because they become much more difficult to control afterwards and can determine the success or failure of the field. Annual weeds are easy to control but are a continual problem. There are two general methods of control available, cultural and chemical. Most successful operators use a combination of the two to control weeds.

Cultural Weed Control

Weeds are easy to control when they are in the seedling stage. Both perennial and annual weeds can be controlled at this stage by simply covering them over with a light cultivation. Most any type of cultivator that fits between the rows will serve this purpose though most growers prefer the rototiller (Figure 9-1). In addition to cultivation, it can be used to cut up and incorporate the canes after pruning and to turn under cover crops.

The root systems of all brambles are shallow and it is important that they are not disturbed during the early part of the season when flower development, fruit set and fruit enlargement are taking place. This is also a critical time for weed control. Cultivate only deep enough and often enough to kill the weeds.

Do not use a rototiller when the soil is wet. It is most effective after the soil has dried enough to crumble when squeezed in the hand. Rototilling heavy, wet soils causes the formation of an artificial hard pan just below the depth of cultivation. This slows water penetration causing it to puddle and run off. Even more importantly, this compact layer impedes air movement into the soil profile and slows the growth and physiological activity of the roots. This is especially detrimental during the critical early summer growth period. Cultivate the field only when necessary to control the weeds. More frequent cultivation "to control moisture loss" is not only expensive but is not necessary. If there are no weeds, don't cultivate.

Control weeds within the row by hand-hoeing or with special cultivators that can be maneuvered into the row between the hills. Even hand-hoeing is difficult in fields where the canes have been allowed to fill in between hills to form a hedgerow and in trailing blackberry fields where the new canes are trained along the ground. Some growers overcome the latter problem by lifting up the canes as they develop and tying them along a low wire on the trellis.

Cultivate frequently enough to control annual weeds and pro-

FIGURE 9-1. The row middles in this planting were clean cultivated with a rototiller to control weeds. (Courtesy of Dr. C. A. Brun)

mote bramble growth during the establishment year. Pull or hoe out perennial weeds as soon as they appear.

Begin to cultivate established fields in the spring as soon as soil conditions are suitable and continue as necessary to control weeds until just before harvest. Some growers use a hammerknife or rotary mower to keep down the weed growth between the rows and do not cultivate. They kill weeds within the rows with chemicals. Traffic through the row middles during harvest keeps most of the weed growth down. Continue cultivation after harvest until early September. Later cultivation tends to keep the brambles growing late into the fall and renders them susceptible to winter injury. Some growers broadcast a rapid growing cover crop such as oats, barley or rye soon after the end of harvest. This helps to control weeds, slows fall growth and hardens off the canes. This practice may not be desirable in situations where the vigor of primocanes is low and the additional time after harvest is necessary to obtain adequate cane size. If this is the situation, you should recognize that the late growth is likely to be injured by unseasonal early freezes.

Sub-Soiling

In established fields, the movement of handpickers and heavy machinery through the field compacts the soil in the row middles. Some growers, especially those who machine harvest or who have plantings on heavy soils, pull a single sub-soiling blade through the middle of each row to a depth of 10-12 in (25-30 cm). Do this during September when the soil is relatively dry. Dry soil fractures best and the roots that are damaged at this time of year have plenty of time to reestablish themselves before the spring growth period.

Chemical Weed Control

Weed control, especially within the rows, is difficult and expensive if attempted solely by cultural means alone. Chemical herbicides are available that are effective for both annual and perennial weed control. They are most often used in combination with cultivation, but some growers have successfully eliminated cultivation altogether. Herbicides are especially useful for controlling perennial weeds.

Herbicides and other chemicals are available for use with brambles but the number and kind are continually changing. Those which are available today for use on cane berries may not be registered for use tomorrow. It is important to keep up-to-date on which ones are registered for use on brambles. Application of an herbicide that is not specifically registered for use on brambles is illegal. Even the application of the chemical at the wrong rate or at the wrong stage of growth may be illegal. The crops for which the materials are registered for use along with information concerning rate and time of application appears on the label. Consult your extension agent or farm advisor for the latest information. Ignorance of the law is no excuse. Illegal use can result in stiff penalties and a ban on the sale of the crop.

Characteristics of Herbicides

There are three general types of herbicides used for brambles: (1) Residual herbicides–those which remain in or on the soil for a long period of time and are picked up by the root systems of weeds, (2) Foliar absorbed herbicides–materials that are absorbed by the foliage of the weeds and translocated throughout the plant, and (3) Contact herbicides–those which kill the parts of the weeds they contact. Each type has special conditions that must be satisfied if weeds are to be controlled without damaging the crop.

Selectivity. The ability of chemicals to kill weeds without damage to the crop is accomplished in a number of different ways. The most obvious, selectivity, results from the fact that different plants have varying degrees of susceptibility to a given chemical. The material can thus be applied at a rate which will kill the weeds without damaging the crop. This difference in susceptibility is often due to the stage of maturity of the crop and of the weeds at the time of application. Mature brambles are more resistant to the action of herbicides than seedling weeds.

Another type of selectivity that is especially useful for annual weed control involves placement of the herbicide where it is absorbed in lethal amounts by the weeds, but not the crop. This is accomplished by applying a low solubility herbicide to the surface of the soil. Rainfall, irrigation, or shallow incorporation dissolves enough of the herbicide to kill the germinating weeds. Because of its low solubility, most of the chemical remains in the surface few inches of the soil away from the roots of the crop plants and continues to kill seedlings over varying periods of time depending on solubility and the rate at which the chemical loses its effectiveness.

Foliar absorbed and contact weed killers are sometimes used selectively by limiting the spray application to the target weeds either by directing or shielding the spray from contact with the crop.

Some herbicides kill grasses and do no harm to broadleaved plants. This type of selectivity is quite common among foliar absorbed materials and is very useful for removing perennial grasses from established brambles. These herbicides can be sprayed on perennial grasses mixed in with the crop. The chemical translocates down to the roots of the grass and kills the entire plant. The reverse of this type of selectivity also exists where the herbicide kills broadleaved weeds and does not harm grasses. These materials are excellent for weed control in lawns but have little use in brambles.

Economical Use of Herbicides

The correct rate of herbicide application is critical for good weed control. Application at lower than recommended rates will usually result in poor weed control and a waste of the material. Higher than required rates damage the crop. It is important, therefore, that the rate of application be calculated correctly and that the sprayer be calibrated to apply the proper amount. Even though a sprayer is calibrated properly, the nozzle openings wear with use and it must be readjusted periodically (Table 9-1).

Since herbicides are so helpful in the management of plantings, it is desirable to learn as much as possible about the herbicides that are available. This includes how they kill weeds, the environmental conditions that favor effectiveness, the types of weeds for which they are useful, and length of time they can be expected to maintain control. It is often possible to adjust or change management practices to improve their effectiveness.

Low solubility residual herbicides require moisture following application to dissolve some of the material so it will be picked up by the weeds. Irrigation of dry soils after application may, therefore, be necessary for good weed control. If dry soil conditions continue too long after application, the weeds may reach a stage at which they will no longer be killed. Heavy rainfall or too much irrigation following application may dissolve the chemical too rapidly causing it to be leached out of the root zone or may cause injury to the crop. This is more apt to occur in light sandy soils. On these soils, growers have learned from experience that, to prevent crop injury, it is necessary to reduce the rate of application below that which is normally recommended. They still obtain good weed control but the length of time the herbicide remains effective is reduced.

Apply the chemical at the proper stage of weed growth. The more mature the weed, the more difficult it is to kill.

Foliar absorbed herbicides must be applied to rapidly growing foliage to be most effective. Slow growing, mature foliage does not absorb and translocate the material satisfactorily. Under such conditions, it is better to either wait until the weeds renew active growth, or to mow them off and force new, more actively growing foliage before application.

TABLE 9-1. Calibration of boom sprayers

Pre-Calibration Check

1. Examine and clean all nozzles and screens.
2. Repair all leaks.
3. Be certain the pressure gauge is working properly.
4 Decide on the gear and throttle setting that will be used for spraying.

Calibration

1. Measure off and mark a test area 100 ft long (30.5 m) with stakes.
2. Fill the sprayer with water.
3. Adjust the pressure gauge and engage the tractor in the proper gear.
4. With the throttle at the correct setting, take a running start and measure the time in seconds required to travel 100 ft (30.5 m).
5. Return the tractor to level ground and with the pump operating at the correct pressure, collect the water delivered by one nozzle over a period of time equal to the time required to travel 100 ft (30.5 m).
6. Determine the volume of water in ounces or millileters.
7. Calculate the application rate using the following formulas:

Formula for Calculating the Gallons Per Acre Application Rate

$$\frac{\text{inches between nozzles}}{12} \times 100 = \text{plot size in square feet}$$

$$\frac{43560}{\text{plot size}} = \text{ratio of plot size to one acre}$$

$$\frac{\text{ounces of water collected}}{128} = \text{gallons collected}$$

$$\text{ratio} \times \text{gal. collected} = \text{gallons per acre application rate}$$

Formula for Calculating the Liters Per Hectare Application Rate

$$\frac{\text{cm between nozzles}}{100} \times 30.5 = \text{plot size in square meters}$$

$$\frac{10,000}{\text{plot size}} = \text{ratio of plot size to one hectare}$$

$$\frac{\text{ml. of water collected}}{1,000} = \text{liters collected}$$

$$\text{ratio} \times \text{liters collected} = \text{liters per hectare application rate}$$

Table 9-1 (continued)

<u>**Example**</u>

A single nozzle used to spray a 20-inch band along the base of plants.

Ounces of water collected = 15

$\frac{20}{12} \times 100 = 167$ sq. ft. plot size

$\frac{43560}{167} = 260$ ratio of plot size to one acre

$\frac{15}{128} = 0.117$ gallons collected

$260 \times .117 = 30$ gal. per acre application rate

Do not apply directed or shielded sprays under hot or windy conditions. Even though the material is normally considered non-volatile, it will still vaporize if it gets hot enough and will cause crop injury. Turbulent windy conditions may cause some of the chemical to be blown onto the crop plant and cause damage.

PRIMOCANE CONTROL

Excess sucker plants develop between and within the rows and act like weeds unless controlled. Either handpull or cultivate them out as they appear.

Primocane Suppression

Vigorously growing primocanes compete with developing flowers and fruit for storage compounds, photosynthates, nutrients, and light during the spring. Temporary suppression of primocane growth during this period releases these growth materials and improves light conditions making them available to the developing floricanes. This can be accomplished either by hand removal, cultivation, or by chemical means. Primocane suppression has become a valuable management tool for growers of red raspberries and trailing blackberries in regions where growth is vigorous. Hand removal in California and chemical removal in the Pacific Northwest, New Zealand, and Britain has been quite successful (Waister, Cormack and Sheets 1977; Crandall, Chamberlain and Garth 1980).

The primocanes are cut off at ground level or killed back with chemical sprays in the spring when they reach a height of 8-10 in (20-25 cm). On vigorous cultivars that produce an abundance of primocanes, a second spray can be applied when the regrowth reaches the same height. New primocanes develop satisfactorily and yields can be increased up to 70%. The process also improves the efficiency of machine harvest.

SOIL MANAGEMENT SYSTEMS

Successful soil management practices control weeds and excess sucker plants, conserve moisture, maintain soil structure, prevent soil erosion, facilitate harvesting and pruning, and produce a good crop of large, high quality fruit.

As was mentioned previously, there are two widely different methods of weed control available, those that rely entirely on cultivation and those that rely on chemicals. There is also an intermediate system that incorporates the advantages of both of these.

Non-Chemical Weed Control

Cultivation is very satisfactory for control of weeds and suckers between the rows. In plantings that are maintained in hills, it can also be used to kill most of the weeds and remove excess canes in the rows. Throw the soil towards the rows during early spring to smother small weeds. Later in the season, pull it back into the row middles. This can be done by hand or with mechanical or hydraulically operated hoes that maneuver in and out of the rows between the hills. Weed control by tillage alone is impossible in hedgerow plantings where primocanes are allowed to develop between the hills to form a continuous row.

At the time of establishment, plant the plants in the bottom of furrows and control weeds by gradually tilling the soil into the furrows as new primocanes develop. Weed control soon becomes difficult because the primocanes of all brambles tend to sprawl on the ground during the first year and it is difficult to cultivate without damaging them.

Mulching

A thick layer of straw or other mulch material spread over the entire planting or just within the rows will control most annual weeds and some perennials. Apply the material about 8-10 in (20-25 cm) thick during the fall or winter following the establishment year. When packed down, it will form a permanent 2-3 in (5-9 cm) thick layer. Add enough new material each year to maintain this thickness. This method of culture controls weeds, maintains soil moisture, and keeps soil temperatures low. It is excellent for home gardeners and small growers, especially those who are interested in organic production or sustained agriculture. The huge amount of material required and the cost of application and maintenance make it uneconomical for large growers.

Chemical Non-Tillage

Chemical non-tillage has been used successfully by some growers (Figure 9-2). It requires large amounts of chemical to maintain adequate weed control and for this reason is not generally recommended. Apply residual herbicides in early spring, or late fall and spring, to control annual weeds between and within the rows. Use foliar absorbed or contact herbicides to kill annual and perennial weeds within the rows. Alternate two or more different herbicides from year to year to prevent the build up of a single chemical to levels that may damage the crop.

Combined Cultivation and Herbicides

A system combining cultivation for controlling weeds between the rows and herbicides for control within the rows is widely used.

FIGURE 9-2. Weeds in the row middles were controlled with chemicals and by mowing. (Courtesy of Dr. C. A. Brun)

It can be used for both hedgerow and hill types of culture during the establishment year and in established fields.

Apply residual herbicides to the surface of the soil at the base of the plants in an 18 in (45 cm) band during the dormant season. Do not apply any more herbicide per square foot (sq. ha) in the band than is used for a broadcast application. In regions where rainfall is heavy and the winters mild, make two applications, one in the fall to control winter annuals and one in the early spring to control summer growth. Use foliar herbicides as needed to control perennial weeds. Control weeds and unwanted suckers between the rows with cultivation or by close mowing. Growers, who machine harvest, often mow the middles with rotary mowers or hammerknife mowers (Figure 9-3).

Limit tillage to a depth of 2-3 in (5-7.5 cm) to prevent damage to the shallow root system of the caneberries. Cultivate just deep

FIGURE 9-3. A hammerknife mower is useful for controlling weeds and suckers in the row middles. It can also be used for cutting up and partially incorporating canes after dormant pruning.

enough and as often as necessary to control the weeds up until the beginning of harvest (Figure 9-4). Discontinue during harvest and make one or two additional cultivations after harvest. Some growers broadcast a fall cover crop just before the last tillage.

Stopping cultivation in the early fall or sowing a fall cover crop helps to slow growth and prepare the planting for winter (Figure 9-5). The cover crop uses up excess nitrogen and moisture and is very effective for slowing growth. Early maturation of the canes makes them more winter hardy and is especially important in regions where canes may be damaged by early winter cold.

Cover crops that are useful for this purpose include oats, barley, winter rye, and annual ryegrass. All of these germinate quickly and grow rapidly during the fall. Oats and barley are usually killed by low winter temperatures. The rye and ryegrass continue growth

FIGURE 9-4. Rototillers are often used for cultivating the row middles.

through the winter and must be cultivated into the soil in the spring before they get too tall to incorporate easily. Frehman (1989) reported that a mat of dead barley suppressed weed growth in the spring and that very little cultivation was necessary.

Permanent Cover Crop Management

A carefully managed permanent cover crop planted between the rows and chemical weed control within the rows has many advantages. It prevents soil erosion, provides good footing for workers, helps prevent soil compaction by heavy equipment, eliminates cultivation and makes it easy to control late fall growth. It is best for medium and heavy soils. Additional nitrogen fertilizer and irrigation are often necessary, especially on light soils. Apply the nitrogen in a narrow band beside the crop row to prevent stimulating growth of the cover crop. Permanent cover crops compete with the

FIGURE 9-5. A fall-planted barley cover crop helps to slow the growth of primocanes and harden them off before cold winter weather occurs. (Courtesy of Dr. C. A. Brun)

brambles for both nutrients and water and unless carefully managed, reduce the size and yield of fruit.

Once established, maintain the cover crop at a low level of fertility. Keep it mowed short during the period from bud break in the spring until the beginning of harvest.

There is probably no single cover that is best for all growing conditions. In British Columbia, Frehman (1989) tested a lawn-type perennial ryegrass, sheep fescue, and white clover. The perennial ryegrass tended to depress growth and yields. Both sheep fescue (*Festuca ovina* L.) and white clover looked promising. There was some invasion of grasses into the clover and it is doubtful that it can be maintained over a long time without periodic renovation and reseeding. A modification of this system allows the natural vegetation to develop between the rows. It is managed the same as if it were a permanent cover.

General recommendations for establishing and maintaining permanent cover crop row middles include:

1. Seed the cover crop in the spring following the first growing season. It can be seeded during later years but there will be a period of one or two years of adjustment during which the roots of the cover and the crop compete for nutrients and water. Yield and growth may be suppressed somewhat during this adjustment time.
2. Use a perennial, nonaggressive grass or legume that is adapted to the local area.
3. Establish a strip between the rows wide enough to support the wheels of the tractor and other heavy equipment.
4. Fertilize the cover crop only enough to maintain it. Place the nitrogen for the berries in a narrow band near the base of the plants.
5. Maintain adequate moisture levels in the soil during the spring and early summer while fruit is forming. Continue to irrigate in regions where rainfall is limited during the summer.
6. Keep the cover crop mowed short during blossom time and while fruit is developing.

SOIL FERTILITY

In addition to water and sunlight, plants require 15 essential nutrients for growth. All of these except carbon (C) are present in the soil and water and are taken up by the plants through the roots. Temperature, soil aeration, nutrient concentration, rate of plant growth, and soil moisture all influence the rate and amount of nutrients taken up. As a result of the different factors that determine the amount of nutrients available to the plants, it is not possible to establish a set formula for the application of fertilizers under all growing conditions. It is, therefore, the responsibility of the individual grower to develop a suitable fertilizer program based upon soil analysis, close observation of plant response, and experience.

Foliar Analysis

Foliar analysis has limited usefulness for monitoring levels of major elements during the growing season. It is much more useful as a supplement to soil analysis for diagnosing nutritional problems associated with mineral deficiencies or toxicities. When growth is poor or visible symptoms of nutrient deficiencies occur, foliar analysis is an excellent means of diagnosing the problem. The final proof is obtained when either foliar sprays or soil applications actually correct the problem (Table 9-2).

Manure

Barnyard manure supplies nutrients and is a good way to build up the organic matter in the soil. If applied in moderate amounts, it is a very useful addition to the overall soil fertility program and helps to build up and maintain good soil structure. It breaks down and releases N slowly during the growing period. This is good for plant growth but the grower has no control over the amount of N released and, if it continues to be released late into the fall, will delay hardening off of the canes and may result in serious winter injury. A fall cover crop will partially overcome this problem. Horse manure may contain unwanted weed seeds.

Apply manure to the row middles in late fall or early winter. Limit the amount of cow or horse manure to not more than 10 T/acre (22 MT/hectare). This is about the amount applied by one pass of a manure spreader. Chicken or pig manure is more concentrated and contains high nitrate levels and should be applied at about half the rate.

All manures are low in P and should be supplemented with about 200 lb/acre (225 kg/ha) of superphosphate. Horse and cow manure often have a high percentage of raw organic matter, such as sawdust or straw, and additional N is necessary to help break it down.

Mineral Fertilizers

The amount of the different nutrients present in the soil varies widely from one region and soil type to another and a comprehensive soil test taken before establishment can provide the information required to develop a good fertility program. It indicates the pH (acidity) and organic matter (an indication of nitrogen levels) and establishes basic levels for phosphorus (P) and potassium (K). It also measures calcium (Ca) and magnesium (Mg) which are used in moderate amounts as well as the minor nutrients (those elements which are essential for plant growth, but used in very small quantities). The more important of the minor elements are boron (B), iron (Fe), zinc (Zn), copper (Cu), manganese (Mn), and sulfur (S). An interpretation of the levels found is made by the soil test laboratory together with recommendations for rates of application. Analytical procedures vary from

TABLE 9-2. Red raspberry foliar tissue analysis standards[1,2]

Percent Dry Weight or Parts Per Million (ppm)

Element	Below Normal	Normal	Above Normal	Toxic
N Nitrogen	2.5%	2.8%	4.0%	4.8%
K Potassium	1.0	1.5	3.0	4.0
P Phosphorus	0.2	0.3	0.6	0.7
Ca Calcium	0.5	0.6	2.5	—
Mg Magnesium	0.2	0.4	1.0	2.0
Mn Manganese	20ppm	80ppm	300ppm	1000ppm
Fe Iron	30	50	150	250
Cu Copper	1	2	50	100
Zn Zinc	13	34	80	300
B Boron	30	46	80	100

1. Foliar analysis standards are for leaf samples taken July 21 through August 10. Collect five of the most recent, fully extended trifoliate leaves plus stems from current season primocanes (usually within 18 in [45 cm] of the tip) from each plant. Select these at random from exposed areas around the edge of the plant. The total sample should include ten representative plants within the planting for a total of 50 trifoliate leaves.

2. From Scheer and Garren 1981.

one laboratory to another so it is important that recommended amounts be linked to the laboratory which made the analysis.

Apply the nutrients that are below recommended rates before planting. This is especially important for phosphorus since it moves very slowly in the soil and must be incorporated into the root zone to make it available to the crop. Broadcast P, K, Mg, Zn and B, if needed, and incorporate them before planting.

Adjust the acidity of the soil during the fall. Raise the acidity of soils with a pH below 5.5 to about pH 6.5 by incorporating agricultural lime. If the Mg level is also low, use dolomitic lime ($CaMg[CO_3]_2$). Lower the pH of soils having levels above pH 7.0 in the same manner using ground sulfur (see Table 7-1).

Do not apply any commercial fertilizer at the time of planting. The roots are very sensitive to fertilizer salts at this time and are likely to be killed. Growers sometimes use a starter solution containing a very weak concentration of nutrients at the time of planting to get the plants off to a good start. After the plants have started to grow, apply a small amount of N fertilizer to the soil surface keeping it several inches away from the base of the plants. No more fertilizer is likely to be needed during the first year.

Beginning with the second year, apply all of the fertilizer in early spring just as new growth begins (Figure 9-6). If N alone, or a combination of N and K is needed, broadcast the fertilizer on the soil surface near the base of the plants. Incorporate fertilizers containing P in one or two bands 12-18 in (30-45 cm) from the center of the row and 3-4 in (7.5-10 cm) deep. There is no advantage to split applications under most circumstances (Crandall 1980) although, on sandy soils where heavy spring rainfall occurs, it may be desirable to divide the N in half and apply the second half at the beginning of bloom.

Foliar Fertilizers

Leaf feeding with one or more of the major elements and minor elements as a routine supplement to the fertilizer program is expensive and unnecessary. There is little evidence that such a program is any better than a good soil fertility program based on soil test results.

FIGURE 9-6. Nitrogen fertilizer is side-dressed in a band close to the base
plants just before growth begins in early spring.

However, some growers may still want to try foliar sprays, espe<
where soil conditions are such that the fertilizer is quickly tied ·
an unavailable form. If so, apply the spray to the new foliage i
spring near the beginning of bloom. Because of the small an
needed, it can often be combined with a pesticide spray.

Deficiencies of minor elements, especially B, Fe, Mg, Mn, and Zn, sometimes occur. These can often be diagnosed by leaf symptoms and confirmed by foliar analysis (Table 9-3). In such instances, foliar sprays are an excellent means of quickly and efficiently correcting the deficiency. The nutrients are taken into the plants quickly and, if the diagnosis was correct, there will be no further symptoms of deficiency on the new growth. Once it has been corrected by foliar sprays, more economical and longer-lasting soil applications of the nutrient can usually be incorporated into the regular soil fertility program to maintain adequate levels.

Nitrogen Fertilizers

Commercial packages of fertilizer list the content of the major nutrients in the container in a series of three numbers on the label. These numbers represent the percentage of each and are listed in the following order: N-P-K. Thus, a bag labeled 16-20-4 contains 16% N, 20% P, and 4% K. Urea N is labeled as 45-0-0. If there are other nutrients present, they are listed in smaller print following the three-letter designation.

Nitrogen is the nutrient used in largest quantities by all of the brambles. If the other essential nutrients and water are present in adequate amounts, the amount of N applied can, within limits, be used to regulate the vigor of the crop. The amount needed to obtain comparable growth varies considerably from one location to another, depending on temperature, soil type, and rainfall. Regions with high average summer temperatures, light soils, and heavy rainfall or irrigation require higher rates. Actual rates of application for commercial plantings of red raspberries and blackberries range from 30 to over 100 lb/A (34-112 kg/ha) of actual nitrogen. Black and purple raspberries require lesser amounts. A new grower should contact his farm advisor or other growers in the area for suggestions. If this information is not forthcoming, start with about 40 lb/A (45 kg/ha) and each year adjust upwards or downwards depending on the amount of growth obtained during the previous year.

Brambles respond equally well to all forms of nitrogen fertilizer. It is, therefore, most economical to use the form that supplies a unit of N at the lowest cost. This can be calculated by using the following formulas:

$$\frac{\text{Cost per Bag}}{\text{Lbs or kg per Bag}} = \text{Unit Fertilizer Cost}$$

$$\frac{\text{Unit Fertilizer Cost}}{\text{\% N in Product}} = \text{Cost per Unit of N}$$

Example: 80 lb bag of fertilizer at $12.00 per bag. Ammonium sulfate–21%; ammonium nitrate–32%; urea–45%.

$$\frac{\$12.00/\text{bag}}{80 \text{ lb/bag}} = \$.15/\text{lb unit cost of fertilizer}$$

$$\frac{\$.15}{.21} = \$.71/\text{lb N ammonium sulfate}$$

$$\frac{\$.15}{.32} = \$.47/\text{lb N ammonium nitrate}$$

$$\frac{\$.15}{.45} = \$.33/\text{lb N urea}$$

Based on the above cost, the urea fertilizer provides nitrogen at the lowest cost per pound.

Effect of Nitrogen Source on Soil Acidity

The portion of ammonium sulfate fertilizer remaining in the soil after the N is taken up by the plants is acidic; after ammonium nitrate, slightly acidic; and after urea, neutral. Calcium nitrate and sodium nitrate leave an alkaline residue. Over relatively long periods of time at high levels of use, each fertilizer alters the pH of the soil accordingly. A grower must, therefore, either change the form of N used, or use periodic applications of lime or sulfur to correct the soil acidity according to the soil test.

Interaction of Pruning with N Fertilizers

Large diameter (over 0.4 in [10 mm]) raspberry canes with close internodes produce the greatest quantity and size of fruit. Also, yields

TABLE 9-3. Symptoms of nutrient deficiency

Nutrient	Deficiency Symptoms	Control
Nitrogen	Older leaves small, yellow-green. Primocanes spindly, short, few in number. Fruit small. Red tips on older leaves.	Increase soil application of N fertilizer.
Phosphorus	Stunted, dark green foliage with purple color on bottom of older leaves. Premature loss of leaves.	Incorporate P fertilizer 3-4 in (8-12 cm) deep in a narrow band.
Potassium	Older, lower leaves small with scorched, brown margins. Young leaves small with yellow-green between veins.	Apply either potassium sulfate or potassium magnesium sulfate.
Calcium	Tipburn of newly developing leaves.	Apply agricultural lime or dolomitic lime.
Magnesium	Older leaves with interveinal reddening, more pronounced near margins. Dead spots later. Stunted growth.	Apply dolomitic limestone or magnesium sulfate fertilizer.
Boron	Delayed bud break, dead buds near primocane tips, fern-like leaves, some dieback of new growth.	Biennial soil applications of Borax at 25 lb/A (27 kg/ha) or Solubor at 12 lb/A (13 kg/ha). Annual Solubor spray in the spring at 5 lb/A (5.6 kg/ha).
Iron	Young, terminal leaves turn white or yellow, leaf margins and interveinal areas turn brown.	Iron chelate sprays whenever symptoms appear. Apply lime and improve soil drainage.
Zinc	Yellow, white or dead between the veins of older leaves, short internodes, rosette-like terminal growth.	Soil application of zinc sulfate at 12 lb/A (13 kg/ha) or summer sprays at 1 lb/A (1.1 kg/ha).
Manganese	Interveinal yellowing of lower leaves, veins and upper leaves normal.	Manganese sulfate 2 lb/A (2.2 kg/ha) or Mn chelate spray during the summer.

increase as the number of such canes increases up to a maximum of about 12-14 per lineal 3 ft (0.9 m) of row (Crandall 1980). Therefore, a cane thinning method that leaves all medium to large diameter canes results in highest yields. This system of thinning canes according to the vigor of the planting is most suitable for wide row spacings and crossarm trellises. In plantings with narrow row spacing, this system causes the canes to be too crowded and limits light.

High levels of nitrogen increase both the number and diameter of the canes produced. If cane thinning is done according to vigor, the advantages of the higher N are fully utilized. In plantings thinned to a fixed number per hill, these advantages are mostly eliminated. It is likely that this interaction between pruning and N levels is responsible for the confusion that exists concerning the advantages of higher N fertilizer levels. In regions where it is difficult to obtain vigorous growth, close row spacing may be necessary to obtain adequate numbers of canes per acre (hectare).

Phosphorus Fertilizers

Though phosphorus is essential to plant growth, additions of P fertilizer, even to soils with low levels of the element present, seldom result in yield increases. It improves root growth and interacts with the other fertilizer elements to increase fruit quality.

Phosphorus dissolves very slowly. For the roots to absorb it, it must be in a location where they can reach it. Apply it in a band along with the N and K in a mixed fertilizer and incorporate it into the soil to a depth of 3-4 in (9-12 cm).

Potassium Fertilizers

Potassium is used in large quantities by brambles. Where the preplant soil test shows K levels to be low, apply annual applications of either potassium sulfate (50% K) or potassium magnesium sulfate (Sul-Po-Mag). Test the soil in the fall every two or three years throughout the life of the planting. These tests will reveal the need for K and also indicate whether potentially dangerous accumulations are developing. Potassium chloride (muriate of potash 60% K) is less expensive than potassium sulfate but there is a danger of chloride injury and it is not generally recommended. Combine the potassium with the N, or N and P in the spring application.

Calcium

Calcium is seldom deficient in the soil. It is only slightly soluble so conditions that slow uptake from the soil such as high humidity or high salt concentration may sometimes cause a temporary deficiency. Use the soil test results to determine need. Agricultural lime and dolomitic lime are the two principal sources of calcium. Use the latter when both calcium and magnesium are needed.

Magnesium

Magnesium deficiency results in reddening of the areas between the veins of older leaves. This reddening is more pronounced near the leaf margins. Excessive levels of K tend to reduce the uptake of Mg. Use magnesium sulfate (Epsom salts), dolomitic lime, or potassium magnesium sulfate (Sul-Po-Mag) to correct the deficiency.

Boron

Boron deficiency causes delayed bud break in the spring and often results in the death of buds on the upper portion of the canes (Figure 9-7). Severe damage is sometimes confused with winter injury.

Apply boron as a soil application or foliar spray. For soil applications, use borax (10% B) at 25-30 lb per acre (27-33 kg/ha) or Solubor, (20% B) at 12-16 lb per acre (13-17 kg/ha). Broadcast it uniformly on the soil surface between the rows. If concentrated in narrow bands, it may be taken up in toxic quantities. A sprayer calibrated for soil applications is a very convenient way to apply the boron.

Foliar sprays provide quick response and are quite convenient. Once the deficiency has been corrected by soil applications, apply an annual spray containing 1.5 lb Solubor per acre (1.7 kg/ha). The best time to apply the spray is in the spring after the plants are in full leaf.

FIGURE 9-7. Boron deficiency causes the upper leaves to curl and become deformed. (Courtesy of Dr. C. A. Brun)

Zinc

Zinc deficiency is often found on, but is not limited to, alkaline soils. Correction of deficiency in alkaline soils is difficult and soil applications alone are seldom sufficient. Spray the canes with zinc sulfate at 12 lb per acre (13 kg/ha) or zinc chelate in the early spring just as the buds are starting to swell. Apply additional sprays during the growing season to correct severe deficiency. Once the symptoms disappear, apply one to three sprays per year to maintain levels.

Iron

Iron deficiency is most often observed on alkaline soils or in waterlogged soils. Lower the pH of the soil with applications of ground sulfur or eliminate excess soil moisture to correct the deficiency. More rapid control can be obtained with foliar sprays of iron chelate but this is only temporary unless the basic cause is corrected.

Manganese

Manganese deficiency is not common. When it occurs, it causes interveinal mottling of the younger leaves. It is sometimes confused with iron chlorosis. Correct it with soil applications of 20-30 lb/A (22-34 kg/ha) of manganese sulfate or spray applications of manganese chelate at 3 lb/A (3.4 kg/ha).

IRRIGATION

Water is essential for cell division and cell enlargement and brambles use large amounts of it during the spring and summer months. Lack of sufficient rainfall during the time of year when rapid growth occurs prevents fruit from developing to its full size and limits the number and diameter of primocanes.

Table 9-4 shows the yield of 'Sumner' red raspberries in a sandy loam soil over a period of eight years. Irrigation was begun in the summer of 1962 using soil sample scheduling. Evaporation pan scheduling was begun in 1965 (Crandall et al. 1969). The latter method was more effective in attaining and maintaining high yields.

Irrigation requires a major investment in capital for equipment

TABLE 9-4. Effect of irrigation on the yield of 'Sumner' red raspberries.[1]

	1961	1962	1963	1964	1965	1966	1967	1968
Yield T/A	2.0	5.9	3.8	4.2	7.3	7.8	6.7	7.3
Yield MT/ha	4.5	13.2	8.5	9.4	16.4	17.5	15.0	16.4

1. No irrigation in 1961. Irrigated in 1962 through 1968. Evaporation pan scheduling of irrigation begun in 1965.

and development of a water source. The decision to develop an irrigation system, therefore, requires much study, especially in temperate regions where summer rains occur more or less frequently and are usually adequate to provide good growth and yield. Many factors must be considered during the decision-making process.

Rainfall and Soil Texture

Nearly all of the moisture used by brambles comes from the upper 4 feet (1.2 m) of the soil profile. This is the region of greatest root development. The available moisture in this part of the soil can be likened to a soil moisture reservoir from which the crop withdraws water for plant growth. As long as there is adequate moisture in it, the plants continue to grow normally. As the moisture is used up by the growth of plants and evaporation, it must be replaced by rainfall, irrigation, or other means. When this does not occur in a timely fashion, the crop fails to obtain enough water and both growth and production suffer.

This 4-foot (1.2 m) soil moisture reservoir has a maximum water holding capacity (Figure 9-8). As water is added to the soil surface in the form of either rain or irrigation, it soaks down through the soil profile, raising the moisture content up to field capacity (the maximum amount of water that the soil will hold against the pull of gravity). The total amount of water present in the soil at field capacity varies depending on the soil texture (the ratio of sand to clay particles) and its organic matter content. This moisture can be removed by plant roots down to the wilting point of the soil (the amount of water left in the soil when plants growing in it wilt and will not recover unless water is added). The amount of water thus available to plants in any given soil type varies considerably.

In regions where summer rainfall is unpredictable, there are occasionally times when the rains are spaced farther apart than normal. If the amount of available water in the soil moisture reservoir is not enough to cover the deficit during this period, the crop suffers.

BRAMBLE PRODUCTION

FIGURE 9-8. Almost all of the moisture used by brambles in a well-drained soil comes from the surface 4 ft (1.2m). This layer of soil is known as the soil moisture reservoir.

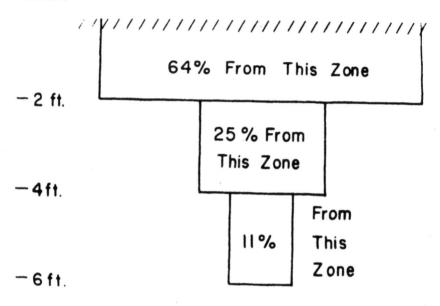

Table 9-5 shows that the amount of water available for plant growth in a loam soil is three times that in a sandy soil. Therefore, a crop growing in a loam soil is much more likely to survive a long time between rains or irrigation than one growing in a sandy soil.

Environment

Nearly all water used by plants is evaporated into the atmosphere through microscopic pores in the leaves called stomates. This method of water loss is called transpiration. Only a very small percentage of the water taken up by the plant is actually used within the plant for growth.

Factors which affect evaporation, therefore, alter the water requirement of the crop. High summer temperature, low humidity, and strong winds all increase water use. Therefore, supplemental

TABLE 9-5. Amount of water available to plant roots in the upper 4 ft (1.2 m) of soils with different textures[1]

SOIL TEXTURE	AVAILABLE WATER[2]	
	Inches	Centimeters
Sand	2.4	6.1
Fine Sand	3.8	9.6
Sandy Loam	5.3	13.4
Loam	7.7	19.5
Silt Loam	8.6	21.9
Clay Loam	9.1	23.2

1. From Crandall et al., 1969.
2. One inch of water is equivalent to the amount of irrigation or rainfall needed to cover the soil surface 1 in (2.5 cm) deep.

irrigation is more likely to be needed in southern climates than in temperate northern regions.

Economic Factors

Irrigation requires a reliable source of good quality water at a reasonable cost. It must be available in sufficient quantity to supply the needs of the crop at critical stages of growth. Even a temporary shortage of water during fruit development will reduce fruit size and quality.

Brambles are sensitive to high salt concentrations (dissolved minerals) in the irrigation water. These are often found in the water of arid regions where irrigation of crops is intensive. Under such conditions, growth becomes stunted and the fruit quality is poor.

Because brambles produce a high value crop, the high cost of irrigation can often be justified. The potential benefits to be derived must, however, be balanced against the initial developmental cost and the cost of operation. When these factors are evaluated carefully, the decision may or may not be to irrigate.

Irrigation Systems

A well designed irrigation system supplies enough water to meet peak demands during the growing season. It does this with minimum interruption of the normal cultivation and harvest operations. Several methods of irrigation are available, each with advantages and disadvantages.

Ditch or Rill Irrigation

This method of irrigation does not require a major investment in equipment but it is not very well suited for bramble production. It has two requirements which are quite limiting. The field must be uniformly level or slightly sloping. This often requires major soil surface shaping to eliminate irregularities before the brambles are planted. The soil must have a rather heavy texture, one that will allow the water to flow freely to the far end of the irrigation ditches before it soaks into the soil. In addition to these limitations, the ditches must be reformed after cultivation and, if they are present during harvest, they create a major problem for pickers.

Form one or more shallow ditches between each row. Divert the water into the ditches from a mainline lateral at the upper end of the field. Allow it to flow to the end of the ditch and then soak into the soil. The uniformity of application depends on whether the water reaches the far end of the field before appreciable amounts soak into the soil. There is a strong tendency for the near end of the field to be over-irrigated. This, combined with the heavier soil texture, is often detrimental to the growth of brambles.

Sprinkler Irrigation

The majority of brambles are irrigated with overhead sprinklers. These range from permanently installed solid set sprinklers or hand-moved sprinklers to mobile units that move down the rows covering up to an acre at a time using a single nozzle. Properly managed, these systems do not increase fruit rot problems. The key to management is to apply irrigation under conditions that allow the foliage to dry quickly. Do not apply it under slow drying conditions or when the application will be followed by an extended period of rainy, humid weather. Such conditions can result in major outbreaks of fruit rot.

A properly designed sprinkler system applies water uniformly over the entire field. Because it is designed to apply it at a specific rate in inches (cm) per hour, it is easy to calculate the length of time necessary to irrigate in order to apply a predetermined amount of water. This precision, combined with a good estimate of the amount of water used by the crop between irrigations, makes it possible to apply water with little danger of either over-irrigation or under-irrigation.

In regions where spring frosts are apt to occur during blossom time, a sprinkler system is often used for frost control. Cold injury occurs at temperatures a few degrees below freezing, depending upon the stage of development of the flowers. Blossoms at or near full bloom are most sensitive and are injured at 29-30°F (−1.7 to −1.1°C). As temperatures drop to near freezing, turn on the sprinklers. Water gives off large quantities of heat as it changes from the liquid state to ice. Its temperature remains at 32°F (0°C) until all of the free water has frozen thus it maintains the temperature of any plant surface it comes in contact with at 32°F (0°C). Leave the sprinklers on the entire time air temperatures are below freezing or turn them on and off at intervals frequent enough to maintain a layer of unfrozen water on the plant surfaces.

A solid set system is expensive to install, but once installed, it is very convenient and requires very little hand labor (Figure 9-9). The main lines and laterals are buried at the time of or before the trellises are installed. Five- or six-foot (1.5-1.8 m) risers are installed for the sprinkler heads. These are located beside a trellis post for support. The system is divided into individual blocks of a size that is appropriate for the capacity of the pump.

A portable, handmoved aluminum pipe system consisting of buried mainlines with portable lengths of pipe and risers for the sprinkler heads is much cheaper than the solid set system, but is very labor intensive. It is quite suitable for medium to small acreages or

FIGURE 9-9. Riser and nozzle used in a solid set, overhead sprinkler system.

in situations where one or two irrigations per season will usually suffice. It also has the advantage that much of the aluminum pipe is available for use with other field crops. The laterals, containing the risers are placed between the rows and, where possible, are moved down the rows from one set to the next. Sometimes, rather than using tall risers for the sprinklers, the trellis posts are modified so that the lines can be placed at right-angles to the rows on top of the posts. Sections of pipe are moved from one set to the next by workers who lift several sections of the pipe above the bramble canopy, often with the aid of pitch forks, and carry them down the row to the next set. This system, including the sprinklers, is readily convertible for use in other crops.

The mobile, single nozzle irrigator consists of a wheeled unit narrow enough to travel between the bramble rows (Figure 9-10). A large nozzle capable of distributing water over a radius of 70 ft (21 m) or more is mounted on the vehicle at a height which clears the tops of the plants. Water is fed to the sprinkler by a large diameter, flexible hose long enough to reach from one alleyway to the next. This hose is coiled on a huge reel mounted on the vehicle. The reel has a variable speed water- or gas-driven motor that moves the sprinkler down the row as it reels in the hose. The speed of the reel determines the amount of water applied per acre (ha). The supply hose is hooked to the hydrant and a tractor is used to pull the sprinkler down the row, unrolling the hose as it goes. At the end of the row, it is unhooked from the tractor and the water turned on. It returns to the head end of the row, rolling up the hose as it goes. It is then ready to be moved to a new location.

When this type of sprinkler was first introduced, it was believed that the force of the large-sized drops necessary for them to be propelled long distances would damage the fruit and knock much of it off of the plants. This has not proved to be true and the system is rapidly becoming popular with large growers because of its convenience and low labor requirement.

Drip Irrigation

Drip irrigation is well-adapted to bramble production. It has a low labor requirement and results in considerable reduction in the amount of water needed for irrigation. There is no effect on fruit rot. The principal disadvantage is the high cost of installation. Also, the plastic lines deteriorate over time or may be damaged by rodent feeding or machinery.

A typical drip irrigation system consists of a pressure regulator and a filter together with the necessary water distribution lines. The

FIGURE 9-10. Single nozzle, mobile irrigator used to irrigate 14 or more rows at one time.

pressure regulator is necessary to insure uniform distribution of the water. The filter eliminates small particles of grit and organic matter that may plug up the emitters. It is either a sand filter or a fine screen filter, both of which have a convenient method for back flushing to clean out accumulated debris.

The mainlines are made of rigid plastic and buried deep enough to escape cultivation. Smaller laterals are installed, one for each crop row. They can be buried, laid on the surface of the soil, or suspended from one of the trellis wires. Buried lines have a longer life span than those exposed to sunlight and weather, but are difficult to inspect for emitter failure. Lines laid on the surface are subject to rodent damage, vandalism, and accidental damage by pruners or cultivation equipment. The system where the laterals are fastened to the trellis wire appear to be the most desirable. They are vulnerable to vandalism and subject to weathering which means that they must be replaced more often than either of the other two methods, but ease of inspection and maintenance is very important.

Many different types of emitters are available. All of them are designed to apply water at a very low rate, usually at 1-2 gallons (3.8-7.6 l) per hour, at a line pressure of about 20 psi. The water is applied to a restricted zone on or near the surface of the soil and distribution occurs both laterally and vertically within the soil profile. The finer the soil texture, the greater the lateral movement of the water.

Two general types of emitters are available, line source and point source. Line source types have the emitter system built into the lateral. They include porous tubes, perforated tubes, or multi-chambered tubes. These are not especially well suited to brambles because they keep the soil around the crown of the plants saturated for extended periods of time and may lead to root rot development. The point source emitters are better. They can be inserted into the laterals at any interval desired. The recommended spacing ranges from 3 to 6 ft (0.9-1.8 m). In established hill system plantings, the emitters are placed between the plants (Figure 9-11).

Emitters may be very simple or quite complicated. The simplest emitters consist of small diameter holes drilled in the lateral at the proper spacing. These are very subject to plugging and the rate of application varies greatly with elevation differences and the length of the lateral. Another simple emitter consists of very small diameter plastic tubing inserted into holes drilled into the laterals. The rate of flow is regulated by the diameter of the tubing and its length. The length of the tubing is varied to regulate the rate of flow. The tubes are wrapped around the lateral to keep them out of the way of pickers and pruning operations. Other more complicated, and more expensive emitters are manufactured to apply a fixed rate or, in some cases, they can be adjusted for rate of flow. Some are self-cleaning or may be easily disassembled for cleaning. Others have built in pressure regulating systems. The more complicated they are, the more expensive they are, though, this additional cost may be offset by lower costs of maintenance and less frequent replacement.

Do not operate drip systems continuously. Vary the frequency and duration according to water use and soil type. In dry, hot locations, use long, daily applications but in areas where there is less moisture stress, reduce the frequency and duration.

WATER MANAGEMENT

Scheduling Irrigation, When and How Much

The actual amount of water used by crops varies from day to day and it is not easy to determine when to irrigate. In areas where rainfall occurs fairly frequently during the growing season, the determination is complicated even more. Even in regions where rainfall is infrequent or entirely lacking during the growing season and growers usually have a fixed irrigation schedule, it is still necessary to know how much water to apply at each irrigation since water use varies with temperature, humidity, wind speed, and stage

FIGURE 9-11. The drip irrigation lateral line is supported by the trellis wire. Emitters like the one shown are most often spaced between every other plant.

of crop growth. Various methods are used to determine the irrigation schedule.

The least complicated method is to watch the plants that are located in the driest part of the field and apply a heavy, soaking irrigation when the first signs of wilting are observed. Unfortunately, by that time the growth of the plants will have already been slowed and production reduced. An improvement on this method involves watching the rainfall pattern during the period prior to harvest. If it appears that moisture is apt to be limited, give the field a good irrigation near the beginning of harvest. With periodic rainfall and soils with good water-holding capacity, no further irrigation is necessary until after harvest. On lighter soils and in areas with less rainfall, apply one or more additional irrigations during harvest to maintain fruit size and growth of the primocanes.

Irrigation on a regular time schedule is common in arid regions. For sandy soils with well drained sub-soils, this does not present much of a problem since water does not build up in the soil and cause root damage. Over-irrigation does, however, leach out nutrients and waste irrigation water. On heavy soils, it presents a much more serious problem. The excess water accumulates and displaces the oxygen from around the soil particles causing the roots to die or to be invaded by root rot organisms.

A more precise method for scheduling irrigation is the soil sample method. Use a shovel or soil sampling tube to extract several samples of soil from the root zone of the crop at a depth of about 1 ft (0.3 m). With experience, it is possible to estimate the moisture of the samples by visual examination and feel (Table 9-6). Samples taken from close to the surface are deceptive since the soil in this area may have been wetted by a light rain while deeper down the soil may still be dry.

Tensiometers and Gypsum Blocks

Another method of determining when to irrigate involves the use of tensiometers or gypsum blocks to monitor soil moisture levels. These devices operate on the principle that as the soil dries, it develops a suction pressure that pulls moisture from inside the tensiometer or block. This suction pressure registers on a gauge and is proportional to the amount of water in the soil. The tensiometer is more accurate in light soils and the gypsum blocks are more accurate in heavy soils. Install them at a depth of 18-24 in (45-60 cm) at several locations throughout the field, the number of locations depending on the variability of the soil. Their readings should represent the extremes in soil moisture as influenced by elevation and soil texture. Monitor them frequently during the irrigation season to schedule irrigation. It is time to irrigate when the readings show that 50% of the available moisture has been used.

There are a few places located near large bodies of water where water is available to the crop from an underground water table. Often, as the season progresses, the water table lowers and this source of water is no longer available. In such situations, install tensiometers or gypsum blocks at a depth of 2-3 ft (0.6-0.9m). Once

TABLE 9-6. Measuring soil moisture by appearance and feel[1]

Soil	AVAILABLE MOISTURE (Percent of Field Capacity)		
Texture	50%	75%	100%
Sand	Appears dry, will not form a ball under pressure	Looks damp, forms a very weak ball under pressure	Forms weak ball, leaves moisture on hand when squeezed, no free water
Sandy Loam	Will not form a weak ball under pressure, looks damp	Forms a weak ball under pressure, breaks easily, not slick when kneaded	Forms a ball under pressure, may become slick when kneaded, no free water
Clay Loam	Forms a ball under pressure, somewhat slick when kneaded	Forms a pliable ball under pressure, quite slick when kneaded	Forms a very pliable ball under pressure, slick when kneaded, no free water

1. Adapted from Harris and Coppock 1978.

the moisture level at this depth drops to 40-50%, it is time to irrigate.

Evaporation Pan Scheduling

The most accurate means of scheduling irrigation is the "check book" method which uses a measurement of evaporation to estimate water use. It gets its name because it is similar to balancing the amount of money spent against the money deposited in a checking account. The amount of water extracted from the soil by the plants over a period of time is estimated with the evaporation pan and enough water is added to make up the difference between water used by the crop and the amount of rainfall that occurs during that interval.

Research has shown that, during the summer, red raspberries extract moisture from the soil profile at the same rate that water evaporates from the free water surface of an evaporation pan (Crandall et al. 1969). The pan used for the research was a 4-foot (1.2 m) diameter, standard U. S. Weather Bureau evaporation pan. For every inch (2.5 cm) of water evaporated from the pan, the raspberries extracted the same amount from the soil. Data are not available for other bramble crops, though it is most likely that their transpiration losses are close enough to those of red raspberries for growers to use the same method to estimate water usage.

One evaporation pan in a county-wide area is enough to serve all growers in the county. All that is necessary is for the individual grower to measure the rainfall on the farm and adjust the evaporation total for the difference in rainfall between the two locations.

Variable Irrigation Schedule

Some growers use a variable irrigation schedule. They wait until the soil reservoir is depleted to 50% and then apply enough water to refill it.

Before this method of scheduling irrigation can be used, it is necessary to determine some basic information about the soil moisture reservoir. In Figure 9-8, the soil moisture extraction pattern of raspberries growing in a well drained, sandy loam soil is shown.

Nearly 90% of the water used came from the surface 4 ft (1.2 m). Therefore, 4 feet is considered to be the depth of the soil moisture reservoir. Root growth in this type of soil is unrestricted. In heavy soils, roots do not penetrate as deep and it is customary to use a 2-foot (0.6 m) reservoir in those soils. The water in the reservoir plus any rainfall that occurs during the interval between irrigations is available for use by the plants. Plant growth is not restricted until after 50-60% of the available water is extracted.

Once the soil type is determined, it is possible to estimate the amount of water available for use between irrigations. Table 9-5 shows the approximate water holding capacity of soils with varying textures. A sandy loam soil holds about 5.5 in (14 cm) of available water in the surface 4 ft (1.2 m) of soil and irrigation is begun when 50% of this available moisture is depleted. Therefore, when the evaporation pan total adjusted for rainfall reaches 2.8 in (7.1 cm), it is time to irrigate with 2.8 in (7.1 cm) of water to bring the soil back to field capacity.

Sprinkler irrigation systems are designed to apply a fixed amount of water per hour. This makes it possible to determine how many hours of irrigation are necessary to apply 2.8 in (7.1 cm) of water. Sprinkler systems do not apply water uniformly, especially during windy weather. For this reason, it is customary to use a 75% application efficiency factor. Applying this efficiency factor to an irrigation system designed to apply 1/3 in (0.8 cm) per hour, it requires 11 hours of irrigation to fill the sandy loam soil back to field capacity.

Refer to the following example:

Water Use between Irrigations (50% Depletion of Reserve)

Effective rooting depth	4 feet (1.2 m)
Avail. water per foot (0.3 m) of soil depth	1.4 in (3.6 cm)
Total water available (4 × 1.4 in)	5.6 in (14 cm)
Use 50% before irrigation	
(.50 × 5.6 in) or (.50 × 14 cm)	2.8 in (7 cm)
Evaporation equivalent	
(2.8 ÷ 1.0)	2.8 in (7 cm)

Irrigation

Sprinkler applic. rate	.33 in (.8 cm) per hour
Sprinkler efficiency	75%
Length of irrigation needed	
$\dfrac{2.8}{.33 \times .75}$ =	11 hours duration

Regular Irrigation Schedule

Some growers irrigate on a regular time schedule. All they need to know is the evaporation pan total for the period between irrigations. The sprinklers are operated long enough to make up this amount.

Water Use between Irrigations (10 Day Interval)

Time between irrigations	10 days
Total evaporation measured for the 10-day period	1.8 in (4.6 cm)
Inches of irrigation needed (1.8 × 1.0)	1.8 in (4.6 cm)
Length of irrigation needed	
$\dfrac{1.8}{.33 \times .75}$ =	7.5 hours

Fixed Number of Hours Per Irrigation

Other growers want to operate their sprinklers on a fixed number of hours each time they irrigate. They need to know how many days to wait between irrigations.

Water Use between Irrigations (Fixed Length of Irrigation)

Desired length of irrigation	8 hrs
Application rate	.33 in/hr (.8 cm)
Application in 8 hours	2.64 in (6.7 cm)

Usable application at
.33 in/hr (2.64 × .75)   2.0 in (5 cm)

Evaporation between irrigations   2.0 in (5 cm)

Irrigate when evaporation shows that 2 inches (5 cm) of water has been used and apply an 8-hour set.

Scheduling irrigation by pan evaporation is relatively simple and very practical. It results in efficient water use with minimum danger of either under- or overwatering. It is desirable that this system of scheduling be monitored for a while after it is begun. If there are any signs of either over- or underwatering, adjust the schedule accordingly.

Chapter 10

Pruning and Training

TRELLIS CONSTRUCTION

Red raspberries and semierect and trailing blackberries require some type of trellis to support their canes and fruit. Install the trellises during the fall or winter prior to the first harvest season. By delaying construction until after the first growing season, cultivation during the first year is easier. Sometimes, if growth is poor and the expected crop does not appear to be economical, mow the canes off during the winter and wait until the following season to install them.

Some trellises consist of a single post set by each plant in the square system of management, but most often they involve posts and wires. The details of construction vary widely (Galletta and Himelrick 1990; Pritts and Handley 1989; Turner and Muir 1985) and depend largely upon the individual grower's evaluation of needs.

Trellises consist of two basic types–the narrow, upright I-trellis and the wider, crossarm or T-trellis. Posts can be made of either wood or steel but the end posts are usually made of wood and are larger in diameter and longer than the posts within the rows. End posts must be anchored or braced well to withstand the strain imposed by the crop and the pull when the wires are tightened. They are usually 8 ft (2.4 m) long and the posts within the row are 7 ft (2.1 m) or longer and spaced 25-30 ft (7.6-9.1 m) apart. Insert the posts into the soil at least 18 in (45 cm). Steel posts can be used in the rows but, in machine harvested fields, they should be spaced 20 ft (6.1 m) apart.

Use 12-gauge (34 ft per lb) (23 m per kg) galvanized wire for the upper support wires and 14-gauge (60 ft per lb) (40 m per kg) for the lower training wires. Heavy yielding, trailing blackberry cultivars ('Thornless Evergreen') require heavier 10-gauge support wires. In regions where growth is less vigorous, lighter gauge wire can be used for both support and training wires. Monofilament plastic wire is sometimes substituted for the metal wire. It is light in weight and easy to handle. It is used in areas where lightning strikes are common because it does not conduct electricity and thus prevents lightning damage to the plants.

Red Raspberry Trellises

Upright Trellises

The upright trellis is required for present-day machine harvesters. It is also used for handpicked trailing and semierect blackberries as well as red raspberries.

This trellis is simple in construction and allows close row spacing. Since many of the primocanes develop on the outside of the rows, they are subject to damage during cultivation, spraying, and harvesting. This damage can be greatly reduced by proper use of training wires. Also, when the primocanes are pulled into the row with training wires, they shade the floricanes and tend to reduce the potential yield. Such reduction in yield is minimal, however, and can be more than compensated for by spacing the rows closer together.

Fasten a single, upper support wire to the posts at a height of 5.0 ft (1.5 m). Tie the floricanes firmly to this wire either singly or in bunches. Scottish growers space the canes uniformly along the wire and tie them to it using a continuous string.

Training wires are used to pull the primocanes into the row during the growing season. This keeps them from interfering with cultivation and harvest. It also helps to prevent mechanical damage to the canes. Two training wires are most often used, one on either side of the post below the support wires. They are fastened temporarily to the end posts and can be adjusted in height by the use of hooks or bent nails placed on each post at heights of 16 and 30 in (40 and 70 cm).

Spread Trellises

The spread trellis has some distinct advantages. The primocanes grow up through the center of the rows where they are protected from damage and require no training. The floricanes are spread out giving better light distribution within the plant canopy and increasing the yield potential. Disadvantages include the need for wider row spacing and the additional cost of trellis construction.

The spread trellis consists of 2×4 in (5×10 cm) crossarms ranging from 18 to 36 in (45 to 91 cm) in length fastened to the top of the posts at a height of about 5 ft (1.5 m). Adjust the height of the crossarms according to the expected height of the primocanes. Training wires are not needed for the longer crossarm trellises since there is adequate space for primocanes to grow up through the center. They are sometimes used for shorter crossarms and are either fastened to the ends of short cross-pieces or the system of adjustable training wires described for the upright trellis is used.

Variations of these spread trellises exist. Some growers use two posts set at an angle of 20-30° forming a V-type trellis. This allows the support wires to be adjusted for height depending on the actual growth of the primocanes. Norwegian growers use the Gjerdes trellis system (Pritts 1989) with adjustable support wires.

Gjerdes Trellis. The Gjerdes system consists of a crossarm trellis with two sets of notches in the arms, one set near the middle and the other near the outer ends. During the dormant season the wires are placed in the middle notches to hold the canes close together. As the fruit laterals develop, they grow outward towards the light. At blossom time, the wires are moved to the outer notches. This allows the primocanes to grow up through the center of the plants and the fruit is concentrated on the outside of the plant canopy where it is easy to spray and harvest.

Trellis for Primocane Fruiting Raspberries

Pritts (1989) also describes a temporary trellis adapted to support primocane fruiting raspberries that are grown for the fall crop only. A hole 3 ft (0.9 m) deep is dug every 25-30 ft (7.6-9.1 m) along the row and lined with a piece of plastic pipe. Midway through the summer, a 7 ft (2.1 m) post with a 36 in (0.9 m) crossarm is placed in each hole. Temporary twine supports are stretched from the ends of the crossarms to hold up the raspberry plants during harvest. After harvest, the twine is cut and the posts are stored until the next season.

PRUNING AND TRAINING RED RASPBERRIES

Training Primocanes

Lay the wires on the ground close to the base of the plants at the beginning of the season. As the primocanes grow, untie the wires from one end post, move them outside of the primocanes, stretch them tightly and then retie them. Hook the wires on the lower of the two hooks. Later, as the primocanes develop, raise them to the upper hooks.

Growers who machine harvest often do not use any training wires. They depend on the machine to pull the primocanes in as it passes along the row.

Removal of Old Floricanes

Floricanes die after harvest and can be removed either at that time or during the winter when cane thinning is done. Removal in summer is sometimes recommended to reduce cane disease infection and eliminate cane borers but experience has shown that weather conditions during this time of year are usually not favorable for cane disease infection. Also, very few cane borers are removed with the canes since most of them are still below the soil surface at this time of year and chemical sprays are very effective for borer control (Scheer et al. 1993). There is, also, an advantage to leaving the old floricanes until winter. If left in place, they help protect the primocanes from wind damage during late summer and fall.

Therefore, removal of the old canes during the summer is optional and depends largely on convenience and the availability of labor. Growers who emphasize sustainable agriculture or organic production usually remove the canes at this time to take advantage of the small amount of disease and borer control that occurs.

Dormant Pruning

Allow the primocanes of red raspberries to grow without interruption through the summer. Except for the optional removal of old canes after harvest, do all of pruning during the dormant season. Use long handled pruning shears or pruning hooks to remove old floricanes that were not cut out after harvest. Then prune off all weak and broken canes at ground level and thin the remaining canes.

Cane Thinning Systems

Two systems are used for cane thinning. In one system, a set number of canes per hill, usually five to eight of the most desirable ones, are allowed to remain. This keeps down the population of floricanes and gives good light distribution within the row. Less crowding may also reduce the severity of fruit rot disease during harvest though this effect is minimal since crowding within the row is largely determined by the number and vigor of primocanes rather than by the floricanes. Some growers use this method of thinning to improve fruit size. Research has shown that, in vigorous fields, this method of thinning reduces potential yield (Crandall 1980) and off-sets the effect of fertilizer applications designed to increase the number and vigor of canes. This effect of fertilizers is discussed in greater detail in Chapter 9 on soil management.

The second method of cane thinning results in variable numbers of canes. Again, all broken canes and those which do not reach the top wire, are removed. Then all vigorous canes up to a maximum of 14-15 per hill or 5 per lineal foot (30 cm) of row are left. The principle behind this procedure is that if the plant is capable of producing this number of vigorous canes, it will also produce the maximum amount of good quality, large diameter fruit.

Topping

Top the canes after they are tied to the trellis and the danger of near 0°F (−18°C) temperatures is past. Remove at least the terminal 6-8 in (15-20 cm) of all canes. This tip portion, if left, produces small, crumbly, poor quality fruit. Cut back the more vigorous canes to about 6 in (15 cm) above the top wire (a maximum cane height of 5.0-5.5 ft [1.5-1.7 m]) depending on the height of the hand pickers. Topping forces strong, fruitful laterals to develop. Top vigorous fields to be machine harvested at 6.0-6.5 ft (1.8-2.0 m) to take advantage of the additional yield that can be expected from longer canes (Figure 10-1).

Disposal of Prunings

Place the canes in the middle between the rows as they are cut (Figure 10-2). After all the pruning is done, cut the canes up with a mower and leave them on the soil surface to decompose, or work them into the soil with a rototiller or disc.

FIGURE 10-1. Canes are topped at 5.5 ft (1.7 m) for handpicking and at 6 ft (1.8 m) or more for machine harvest. (Courtesy of Dr. C. A. Brun)

FIGURE 10-2. Prunings are gathered in the middle of rows and either cut up and left on the soil surface to disintegrate or incorporated into the soil. (Courtesy of Dr. C. A. Brun)

Training Red Raspberry Floricanes

The canes are then ready to be fastened to the trellis. They are either tied individually to the support wires or in bundles. Training systems which space the canes out along the row tend to give better light distribution and higher yields, however, they are more labor intensive.

Linear Systems

Individual tying is suitable for small numbers of canes with the upright, two-wire trellis. It is commonly used in Scotland. They use a single length of twine to tie the canes, making an overhand knot around each as they space them uniformly down the row. A similar system utilizes two support wires. The canes are distributed uniformly between the two wires which are then tied or clipped tightly together at frequent intervals (Figure 10-3).

Bundle Systems

Bundle systems of fastening the canes to the wires are suitable for low to medium numbers of canes. The simplest system is to gather all of the canes from a hill together and tie them in a single bundle to the top wire. A second tie is made about half way down the plant. The tepee system involves separating the canes into two bundles. The tops of adjacent hills are tied together on the support wire (Figure 10-4). In another system, the canes are divided into four bundles and two are tied to each wire of a divided trellis.

T-Trellis Systems

Larger numbers of canes can be accommodated by using a spread trellis. The canes are divided evenly between the two wires. They can be distributed uniformly along the wires and tied individually as in the Gjerde moveable wire system (Pritts 1989), tied in bundles of three to four canes or spaced between two support wires at the ends of the crossarms.

Weaving. Another training method that is suitable for vigorous fields with long canes, involves weaving the canes on the wires (Figure 10-5). Half of the canes are woven on each wire of a spread trellis. One to three canes are brought up inside of the wire and bent over to the outside and down. By slanting the canes down the row, this bend can be made without breaking the canes. The ends of these canes are tucked inside of the next bundle of one to three canes which holds them in place without tying. The ends of the canes are broken or cut off after weaving to eliminate the small, crumbly fruit associated with the weak tips. The yield potential is high and the fruit is concentrated on the outside of the plants where it is easy to pick. The primocanes grow up through the middle without any need for training. There is, however, a tendency for fruit rot to be more serious during wet weather and pickers sometimes damage fruit laterals trying to reach fruit on the inside of the plants.

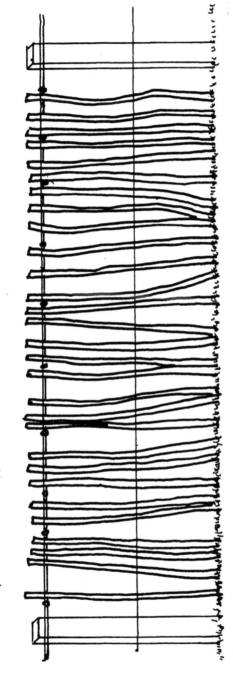

FIGURE 10-3. Red raspberries trained in the linear system. The canes are either tied individually along a single support wire or clamped between two support wires.

FIGURE 10-4. Red raspberries trained by the tepee, bundle method.

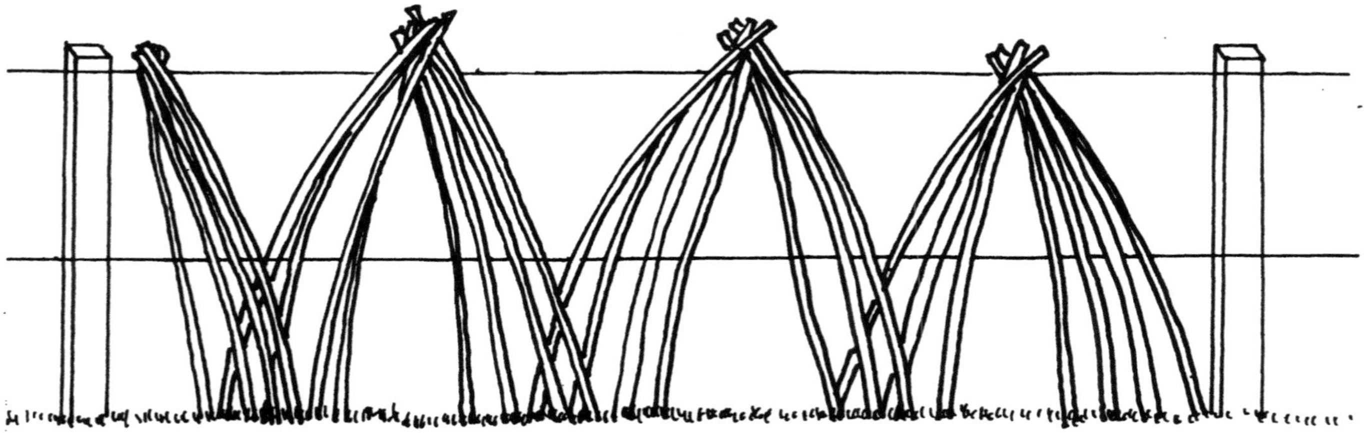

FIGURE 10-5. Red raspberries woven on a two-wire spread trellis.

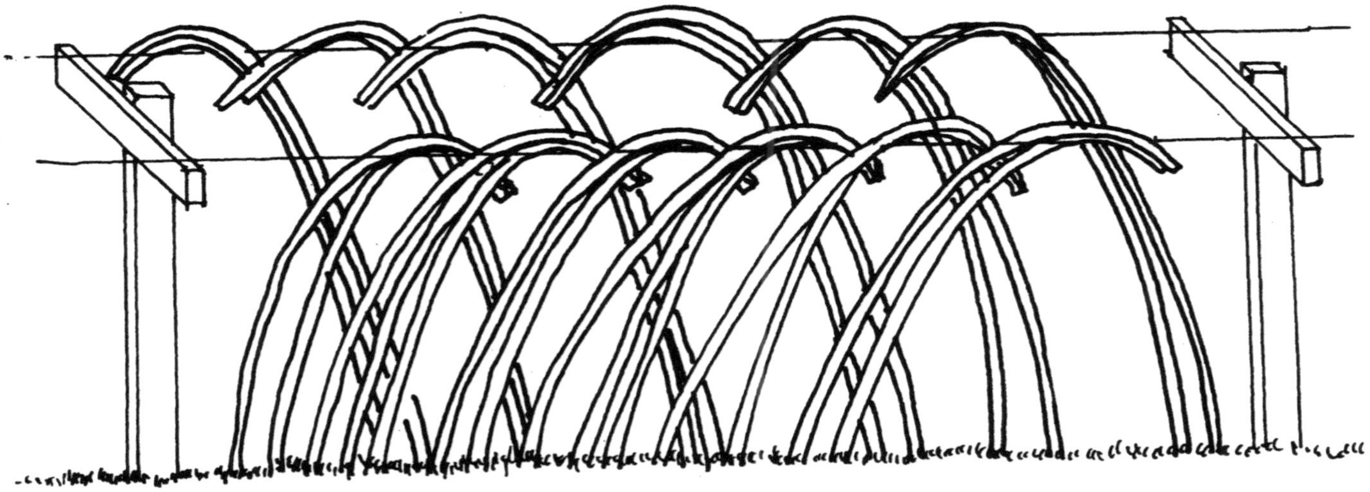

Pruning Primocane Fruiting Red Raspberries

Primocane fruiting cultivars flower and produce fruit on the terminal 1/3-1/2 of the primocanes during late summer. The lower portion of the canes produces a crop the following spring. In the meantime, new primocanes grow that will produce a fall crop. Many home gardeners handle their everbearers this way to give a summer-long harvest.

Most commercial growers use everbearers to produce a fall crop only. They cut off all canes at ground level during the winter. This eliminates the early summer crop but allows the plants to produce a large, good quality fall crop. They use better quality, heavy producing summer cultivars to produce fruit during the early part of the season.

This system for handling everbearers has many advantages. The extensive labor used for dormant thinning, topping, and tying is eliminated. Winter pruning consists only of mowing off the canes as close to the ground as possible and this process can be mechanized. No permanent trellis is required, but if the plants become top-heavy with fruit, a temporary trellis is sometimes erected for support. This system makes it possible to produce a good fall crop in more northern climates where severe winter temperatures often kill or damage overwintering floricanes.

PRUNING BLACK AND PURPLE RASPBERRIES

Summer Pruning

Black raspberry primocanes are always topped during the summer. This causes them to develop strong, productive lateral branches. When the primocanes reach a height of 24-28 in (60-71 cm), remove about 4 in (10 cm) of the terminal growth by hand-pinching or with pruning shears. Go over the planting several times at intervals of about two weeks to top all of the primocanes at the same height. Yields of vigorous cultivars can be increased by higher topping but such plants usually require a trellis for support.

Purple raspberries, too, can be tipped during the summer, but they can also be left untipped and handled the same as red raspberries. They are more vigorous than black raspberries and are therefore, tipped about 8-12 in (20-30 cm) higher. Those plants which are topped at the lower height do not need to be trellised for support. Although this lower height reduces potential yield, the loss is more than compensated for by the elimination of the cost of trellis construction. In some regions where heavy, icy snow occurs, the plants require a trellis anyway to prevent the weight of the ice and snow from breaking off the laterals.

Black and purple raspberries are quite susceptible to cane diseases, especially in warm, humid, rainy climates. Under such growing conditions, remove the old canes after harvest. With less severe disease infection, the old canes can be left until winter to provide support for the primocane laterals and prevent damage from wind and weather.

Dormant Pruning

Wait until after the most severe winter weather has passed before dormant pruning. This lets you cut out cold-injured wood and leaves fewer sites for cane disease infection. This delay may not be economically feasible with large plantings and may not be necessary in more temperate climates.

Cut out old floricanes if they were not removed after harvest and remove all dead wood and weak canes leaving five or six strong floricanes per hill. Since the largest fruit is produced on the basal portion of the strongest laterals, head the laterals back in proportion to their vigor. Eliminate cold injured wood and either remove weak laterals completely or cut them back to 3-4 in (8-10 cm). Remove the laterals that are too close to the ground. Cut back large diameter laterals to 14-18 in (30-45 cm).

Some large growers who produce fruit for processing have completely mechanized the pruning and harvesting operations. They use power hedgers to tip all primocanes at about 20 in (50 cm) immediately after harvest. During the winter, the same hedgers are used to mow the tops and sides to a length of 24 in (60 cm). Old floricanes are left in the row to help support the plants. This method of pruning results in low yields and the life expectancy of the planting is reduced, but the entire operation is mechanized, thus costs of production are very low and net profits are usually satisfactory. The fruit thus harvested is not suitable for fresh market use but can be used for processing.

Purple raspberries are pruned during late winter depending on how they were handled during the summer. Those which were left unpinched are pruned and tied to the trellis the same way red raspberries are handled. Plants that were summer pinched to induce branching are pruned like black raspberries except that the laterals are left 2-3 in (5-9 cm) longer.

PRUNING AND TRAINING BLACKBERRIES

Blackberry Trellises

Erect and semierect blackberries require some type of support for their canes. Often, this consists of an upright, post and wire trellis with two or three support wires depending on how the canes are to be trained. The trellises used for blackberries require heavier posts spaced closer together and stronger support wires because more of the weight of the fruit is borne by the trellis and wires. Most trellises are of the narrow upright type, however, some training systems require cross-arms.

Erect Blackberries

With proper pruning, erect blackberries do not require a trellis for support. During the first year after planting, they grow prostrate and many growers become convinced that they will need a trellis. However, they develop thick, upright primocanes during the second season and do not require trellises. Some thornless types require two years to develop this fully upright growth habit. During the first year, train the developing primocanes along the row on the ground close to the base of the plants. This keeps them out of the way of cultivation and spray operations. Allow the sucker plants to develop in the row and fill in the spaces between plants to produce a continuous hedgerow. Remove all sucker plants that grow between the rows. Prune off all canes at soil level at the end of the first growing season.

Some home gardeners and the commercial growers in tropical regions who harvest the fruit for air shipment, use trellis supports and allow the canes to grow to 5-6 in (12-15 cm) above the top trellis wire before they tip the canes. This provides more economical use of space for home gardeners and better access to the fruit for the fresh market growers.

Summer Pruning After the First Year

When the primocanes reach a height of 3-4 ft (0.9-1.2 m), pinch or cut off 3-4 in (8-10 cm) of their tips to induce branching. Since they do not all develop at the same rate, go over the field several times. Some growers tip the shoots at 3 ft (0.9 m) the first time over the field and increase the height of each later tipping up to a maximum height of 4 ft (1.2 m) (Lipe 1986). This gives a better distribution of fruit laterals and higher yields.

Cut out old fruiting canes after harvest or during the winter. The time of removal depends upon which renders the most efficient utilization of labor. Remove excess suckers that develop between and within the rows.

Dormant Pruning

In northern areas where blackberries are often injured by cold winter temperatures, do not prune the laterals until late winter. At that time, there is less likelihood of further winter injury and, if damage has occurred, the damaged wood is visible and can be removed when pruning.

Remove old floricanes along with any weak or insect-infested canes. Thin floricanes to 4-6 per lineal foot (30 cm) and cut back the laterals to 12-18 in (30-45 cm).

Pruning Trailing and Semierect Blackberries

Trailing and semierect blackberries seldom produce suckers from root buds. All of the new growth, therefore, arises from crown buds at the base of the plants and it is not necessary to hoe out excess sucker plants during the summer.

The primocanes of trailing types are long and slender. As they develop, train half of them in each direction parallel to the row and as close as possible to the base of the plants to keep them out of the way of cultivation. No summer pruning is required other than the removal of the old floricanes after harvest. Many growers wait until winter to cut out the old canes. During the summer, it is difficult for untrained workers to determine which are the old canes so some growers who wish to remove the canes after harvest spray the base of the floricanes in the spring before growth starts with a water-based paint. This eliminates the uncertainty and speeds pruning.

Dormant Pruning and Training

In colder climates, leave the canes on the ground until late winter. Sometimes the canes of trailing types are covered with straw or soil to help protect them from cold injury. Training should be finished before the buds begin to swell. At that stage of development, bud loss from handling is excessive.

Lift up eight to ten of the most vigorous canes and train them on the trellis. Prune off the weaker canes and the old floricanes.

Training Trailing and Semierect Blackberries

Training Primocanes During the Summer

Whereas the primocanes of raspberries require very little summer training, those of semierect and trailing blackberries need considerable attention during the summer because of their prostrate, vigorous growth. This can be accomplished in various ways depending on the cultivars involved and the desire of the grower.

Training on the Ground. The primocanes are often trained on the ground. Divide the developing canes into two halves and train the halves in opposite directions from the base of the plant. Use 2-ft (0.6 m) long stakes to push them into the row and hold them out of the way of cultivation, spraying, and harvesting. This system is not particularly good for semierect cultivars because of their natural tendency to produce upright growth but it is well suited for trailing types. In cold climates, the position of the canes on the ground helps to protect them from the cold and it is easy to provide additional protection with a cover of straw or other mulch material. The canes are susceptible to infection by disease spores washing onto them from the floricanes. During inclement winter weather, lifting them up and training them on the trellis is a disagreeable job and some breakage always occurs.

Training Primocanes on the Top Trellis Wire. A system better adapted to semierect cultivars involves tying the primocanes together loosely as they develop and training them up through the plant to a top trellis wire. There, they can be divided and tied to the wire. This system takes advantage of the natural upright growth tendency of semierect blackberries and positions the canes away from potential disease infection. It can be used for trailing types but is more labor intensive and requires an additional upper training wire. The canes are very susceptible to winter injury in cold climates.

Training Floricanes

Train floricanes either in late summer or during the dormant season. In mild climates, late summer is best (Bullock 1963). At that time, the canes are flexible and easy to handle making them less subject to injury. Experience has also proved that plants trained at this time of year, provided the canes are not injured by cold winter temperatures, produce higher yields. In cold climates, the canes should not be trained on the trellis until late in the winter when the danger of severe cold temperatures is past.

The method of training on the trellis varies. When a single post is used for support, the canes are merely gathered up and tied to the post. They are then cut off at a height of 5 ft (1.5 m). Other systems are designed to retain more of the canes' length and distribute them for maximum exposure to sunlight.

Weaving Floricanes on Two-Wire Trellises. The two-wire, upright trellis is widely used for trailing blackberries and is considered necessary for today's machine harvesters (Figure 10-6). It is inexpensive and very well adapted to vigorous, high yielding cultivars.

Lift the longest canes up over the top wire and under the bottom wire in a spiral. Space additional canes parallel to them until all of the canes have been trained. Those too short to spiral can be tucked

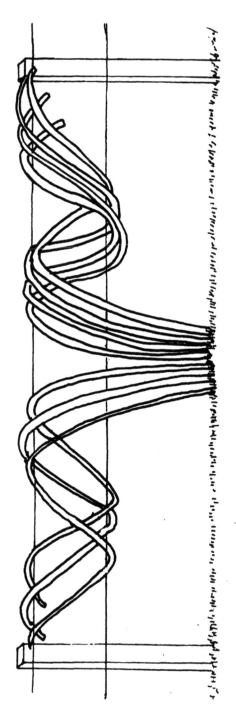

FIGURE 10-6. Blackberries woven on a two-wire trellis.

in among the older canes. After tying, cut the canes off half way between plants so they do not overlap.

Training Floricanes in a Fan Shape. The fan shape training system requires a two-wire trellis (Figure 10-7). It is especially well suited for semierect cultivars and for less vigorous trailing types. Plants can be spaced close together and the canes are distributed uniformly over the whole trellis providing maximum light exposure and yield. It requires more labor and is more costly than weaving.

Alternate Side Training. Like its name suggests, alternate side training involves training the primocanes on one half of the trellis and the floricanes on the other half. Each year the position of the primocanes and floricanes is reversed on the trellis. This system allows the primocanes to be fastened in a permanent location on the trellis during the time of year when the least amount of damage will be caused by handling them. It requires that they be trained and tied during the busy summer months and essentially limits production because of the reduced amount of space on the trellis available for training the canes.

Pruning and Training Erect Blackberries in the Subtropics

Erect blackberries are grown in the subtropics for air shipment to markets in more temperate climates. Under the temperature and day-length conditions in this region, cultivars with a low chilling requirement flower and produce fruit throughout the year and the canes tend to be perennial. The object of pruning is to maximize fruiting during the winter and spring and eliminate it altogether during the rainy summer months. This provides fruit for shipment during the time of year when no blackberries are being produced in more northern markets. Renewal of the planting during early summer eliminates diseased canes and bypasses the rainy time of the year when fruit rot is most apt to occur.

Use two-wire trellises spaced about 6.5 ft (2 m) apart to support the plants. This spacing requires either hand labor or special narrow tractor equipment for cultivation and spraying but the fruit is easily accessible for harvest.

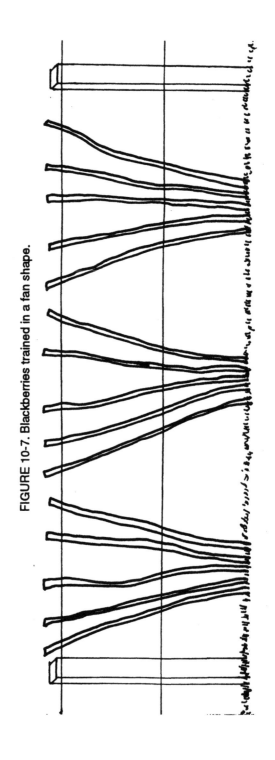

FIGURE 10-7. Blackberries trained in a fan shape.

First Growing Season

The best time to establish a new planting is during the early spring. This provides good moisture and temperature conditions for growth. Allow the primocanes to trail on the ground and train them parallel to the row close to the base of the plants until they are long enough to reach the top wire. As soon as they are long enough, space five or six canes per plant along the top wire and tie them to it. Top them 5-6 in (12-15 cm) above the wire. Remove all excess canes and plants that develop around the base and between plants. When handled in this manner, blackberries will produce fruit during the first winter and spring after planting.

The Mature Planting

Mow off all canes close to the ground during late May or early June. At the same time, spray to control cane diseases and fertilize and cultivate to promote rapid growth. As was done in the first year, tie four to six primocanes per plant to the top wire and top them.

Many strong laterals develop over the entire length of the canes after topping. These can be classed into three types of development: (1) Rapidly growing vegetative laterals that have a slender region extending back about 1 in (2.5 cm) from the growing tip. These laterals are sometimes called "long necks." The terminal leaves have five leaflets. (2) Laterals that no longer have the slender growth region and appear to have stopped growing. The terminal leaves usually have three leaflets. These laterals have initiated flower buds but none of the buds are visible. (3) Laterals with visible flower buds at the tip and in the axils of lateral leaves.

The first laterals that develop are vegetative. Remove all of those that arise from the base of the plants below a height of 14 in (35 cm). When the others reach a length of 15-20 in (38-50 cm), cut them back to 8-10 in (20-25 cm). Go over the field every ten to 14 days as they develop. They will produce new secondary laterals, some of which will have progressed to stages two and three and should be allowed to produce fruit. Continue to head back all vegetative, "long-neck," laterals when they reach a length of 15-20 in (38-50 cm). This continual heading back is necessary to keep the plant canopy from becoming too wide and interfering with harvest and cultivation. Heading back also encourages them to become fruitful.

During the harvest season, the terminal portions of laterals produce fruit and die back to a point just below the last fruit to ripen. Cut them back to this point. This causes new laterals to develop, each of which will bear fruit and help maintain production.

Chapter 11

Propagation

Healthy, vigorous, true-to-name planting stock is essential for success. Fortunately, it is possible to obtain plants which meet these specifications at a reasonable price by purchasing them from a reliable nursery. Although officially certified disease-free plants are not always available, most major nurseries practice propagation methods that insure the health of the plants they sell. Several different methods of propagation are used to produce the stock.

PRODUCTION OF CERTIFIED PLANTS

Most reliable nurseries have developed procedures to insure disease-free plants (Achmet et al. 1980). They begin with a small number of true-to-name plants that have been tested in the laboratory and proved to be free of diseases and viruses. These are held in greenhouses or in isolation to prevent transfer of viruses into them by aphids from outside sources. These "nuclear" stocks are used as propagation material for establishing plant production fields.

Propagation fields are located at least a half mile from any possible source of virus infection. The soil must be disease-free. It is usually fumigated ahead of time to eliminate nematodes and provide further insurance against soil diseases. During the time the plants are developing, the fields are sprayed with insecticides to control aphids and inspected numerous times for the presence of abnormal plants. Such plants, if found, are dug up and destroyed. These procedures result in virus-free plants with healthy root systems. They are sold as certified plants.

Tissue Culture Plants

Another source of certified plants is by tissue culture (Donnelly and Daubeny 1986; McPheeters, Skirvin and Hall 1988; Snir 1988). The growing tips of virus-tested plants are removed and placed under sterile conditions on a nutrient media in sealed test tubes. With the right nutrient media mixture, the cells multiply and form small plantlets. These plantlets are removed and planted in flats of sterile soil and grown in the greenhouse. They are sold as rooted plants in a small ball of soil. Although more expensive than field-grown plants, they are becoming quite popular since they serve as a rapid method for increasing new cultivars and can be readily transplanted by machine.

PROPAGATION PROCEDURES

Plants are obtained from suckers that arise from root buds, from tip layers, root cuttings, or tissue culture. Not all of these methods are suitable for all brambles.

Red Raspberries

Most red raspberries are propagated as rooted suckers that arise from root buds. Suckers from healthy plants are spaced 3-4 ft (0.9-1.2 m) apart in disease-free soil. They are grown for one or two years during which time numerous sucker plants develop. All of the plants are dug during the following winter. Their tops are trimmed back to about 8 in (20 cm) and they are sorted into grades, tied into bundles of 50 or 100, and placed in corrugated boxes holding 500 or 1,000 plants each. If they are to be held for some time before planting, the boxes are lined with polyethylene film. They can then be held in cold storage at temperatures of 28-32°F (7-8°C) for several months (Figure 11-1).

Red raspberries can also be grown from root cuttings (Torre and Barritt 1979). This procedure is not commonly used, but in years when planting stock is in short supply, it provides a good source of plants. One to 2 oz (28-56 g) of unsorted, random diameter and

FIGURE 11-1. A box of certified plants ready for shipment. When the container is lined with polyethylene film, dormant plants can be held in cold storage at 32°F (0°C) for several months.

length cuttings are used per 36 in (0.9 m) of row. They are spaced uniformly along the row to form a hedgerow or at intervals to form hills.

Black and Purple Raspberries

Black raspberries and most purple types are propagated from tip layers. In the late summer and fall, the growing tips of primocanes and laterals bend over towards the soil surface and develop a slender, light-colored, "rat-tail" appearance with small, curled leaves. These tips form roots readily when they touch the ground, though the usual practice is to make a slot in the soil with a spade and insert the tip of the plant vertically into the slot. These tips form sturdy plants. During late winter, they are cut loose from the canes and dug for transplanting. Some cultivars of purple raspberries also form root suckers. These can be dug along with the tip-layered plants.

Blackberries

Erect blackberries reproduce readily from either root suckers or root cuttings. The root suckers are dug during late winter and handled the same as red raspberry plants. Root cuttings are just as successful as sucker plants (Moore et al. 1978) and provide a more rapid method of propagation for either nurseries or commercial fields.

Root Cuttings

The roots of healthy plants are dug and sorted into those ranging in diameter from 1/4 to 3/8 in (6-9 mm). Larger roots are not suitable because the root buds are spaced farther apart and the cuttings may not have any viable buds present. The roots are cut into 3-4 in (7-10 cm) lengths. They can then be heeled-in in well-drained soil or placed in plastic bags and stored in cold storage at temperatures of 32-36°F (0-2°C). Sometimes growers start single root cuttings in plastic bags filled with a well-drained soil mix. The plants thus produced can be transplanted directly to the field.

Trailing Blackberries

Trailing blackberries do not reproduce readily from root suckers and, in the case of chimeral thornless cultivars, such plants would be thorny. They can be propagated by tip layering, stem cuttings, or by tissue culture. The tip layers are handled the same as described for black and purple raspberries.

Stem cuttings are propagated from healthy primocanes which are cut into two- or three-leaf sections. These cuttings are planted in porous soil and rooted under mist in a propagation bed. Once rooted, they are transplanted to the nursery area and held until needed.

Tissue culture can be used for all trailing blackberries including the chimeral thornless types (Caldwell 1984). This method of propagation is especially useful for increasing the stock of new cultivars of which there is limited propagation material.

Chapter 12

Fruit Production

Seldom, if ever, is the true potential yield of brambles attained. Somewhere, some way, yields don't meet expectations. What can you do about it? This chapter is an attempt to provide some of the answers. As with any mystery, the solution involves recognizing the clues and fitting them into their proper place in the puzzle. It is then up to you to take the necessary steps to arrive at a solution.

Evaluate your planting every year at the end of the growing season. Relate your findings to observations made during the spring and summer. Are you satisfied? What can you change to improve profitability and make the operation more enjoyable?

Some reduction in yield may be intentional as in the case of growers who choose to machine harvest. They are willing to sacrifice up to 15-20% in yield in order to eliminate problems and costs associated with handpicking.

Yields are often reduced by climatic or soil conditions associated with the particular site chosen for the planting. The reduction can also be caused by the interaction of other factors, each of which individually may have a minor effect but when combined, reduce yields substantially.

CLIMATIC FACTORS

Reducing the Effects of Summer Heat

Select cultivars with a known ability to grow in hot climates. Blackberries withstand the heat better than raspberries but they, too, are adversely affected by hot, dry weather (Figure 12-1).

FIGURE 12-1. Fruit exposed to the hot, dry summer sun is often sunburned on the exposed side. (Courtesy of Dr. C. A. Brun)

Irrigation

Schedule irrigation very carefully. The method described in Chapter 9 using an evaporation pan to estimate water use is simple and very accurate. Plants use water very rapidly during hot weather and are likely to suffer temporary shortages when scheduling is done by other, less accurate means.

Cool the plants by operating the irrigation sprinklers during harvest through the hottest part of the day. Solid set systems are especially useful for this purpose. Keep the rate of water application low so that the soil does not become waterlogged. Stop the application early enough in the day that the foliage has a chance to dry before nightfall. Use this evaporative cooling only during the harvest season.

Training Systems

Orient the rows in a north-south or northeast-southwest direction. Use an upright trellis system for raspberries and do not pull the primocanes tightly into the row when using the training wires. For trailing and semierect blackberries, use an overhead primocane training system. Take full advantage of the shade provided by the primocane leaves to prevent sunscald of exposed fruit.

Windbreaks

In regions where hot dry winds occur, plant windbreaks to filter and slow the speed of the wind. The object is not to stop the wind completely, but to reduce its velocity and still retain some air circulation within the field. Lack of adequate air movement during the summer causes the humidity to increase and favors the development of fungus diseases. During the spring, cold air settles in low areas and kills blossoms. For these reasons, deciduous plants are better than evergreen plants for the windbreak.

Prevention of Cold Injury

Select cultivars suitable for the climate. Slow the growth of primocanes in the fall to harden them off before severe winter temperatures occur. Plant an annual cover crop in late summer after harvest. Stop irrigation and cultivation by early September. In regions of very severe winter temperatures where cane injury is common, choose primocane fruiting cultivars and grow them for the fall crop only.

MANAGEMENT PRACTICES

Overcoming Heavy Soils

Install subsoil tiles to drain away excess water. Irrigate only as needed. Design the irrigation system to apply water slowly enough that all of it soaks into the soil soon after application. Plant on raised beds so the surface water drains away from the rows. Use green manure cover crops before planting to raise the organic matter content of the soil. Apply annual applications of manure. Pull a subsoiler tine through the row middles during early September each fall to break up any hard pan that may have been created by machinery traffic.

Cultivar Selection

Keep up to date on the latest cultivars available for your area so when the time comes to renew a planting, you will have the necessary information to make an intelligent decision. Be certain that the cultivar you choose is the best for the market you wish to serve. Large-sized, disease-resistant, thorn-free, very productive cultivars specially adapted to the fresh and processing markets are being developed by the breeders and present tantalizing possibilities for new plantings.

Cane Number and Quality

Number and quality of floricanes produced per unit area are directly related to yield. Cane number can be increased by reducing the distance between rows or by increasing the number of canes per hill or per unit of row length when pruning.

Row Spacing

Upright trellises allow close row spacing. The distance selected, though, depends primarily upon the width of the machinery, the vigor of the cultivar, and its growth habit. Once the field is planted, there is little that can be done about row spacing. It should, however, be considered for new plantings.

The Gjerdes Trellis

For those who are willing to test new ideas, try the Gjerdes training system described in Chapter 10. It has real possibilities for improving the ease and efficiency of handpicking and for improving disease and insect control. Any crossarm trellis can be converted to this system with minimum effort and cost. The movement of the training wires from the center position to the outside at blossomtime is the only major change involved.

Plant Stand

As a field ages, plants are lost or become weakened. Eventually their number reaches a point where consideration should be given to replacing the field. If 15-20% of the plants are weak or missing, it is probably time. Before doing so, look the field over to determine why the plants are sick or missing. Sometimes it is possible to correct what is wrong and bring the weak plants back to health. If a new planting is deemed necessary, can you correct these problems in the present location, or should you look for a more favorable site? In either case, make every effort to eliminate all problems in the new planting.

Disease Problems

Wet soil root rots can be alleviated by improving the soil drainage. In new plantings, it may be possible to move to a better drained site. If not, much can be accomplished by installing a well-designed sub-soil drainage system. Phytophthora root rot can be held in check with annual soil applications of fungicide (Scheer et al. 1993). Preplant fumigation plus the fungicide is also very successful. Other diseases that affect cane number and quality are root or crown gall, verticillium wilt, anthracnose, and virus diseases. Anthracnose can be held in check with an annual application of sulfur fungicide at the green bud stage in the spring and by pruning out diseased wood. The other diseases can be prevented by choosing a disease-free site and using certified planting stock. Reuse of the same site requires preplant soil fumigation plus certified plants.

Nematodes

The role of nematodes in yield reduction is somewhat confused. There is no doubt that the dagger nematode (*Xiphinema* spp.) transmits tomato ringspot virus and is serious if the virus is present in the field.

Damage caused by other nematodes is less clear cut. The general consensus is that high populations reduce plant vigor especially in combination with root rot organisms. Test the soil for the presence of nematodes before planting and fumigate only if necessary.

Insect Problems

Look for crown borer damage. Holes in the center of the stubs when dormant pruning and frass or sawdust on the crowns are symptoms of their presence. There is a very good insecticidal spray program for their control (Scheer et al. 1993).

The effect of other insects on growth and production varies widely with the pest and size of the population. Learn to recognize the symptoms of their feeding and take appropriate control measures when they become economically important. Insecticides pose a potential residue threat and are expensive. Use them only when necessary.

Herbicide Damage

Maintain perennial weed control within the plant rows. Grassy weeds seriously reduce the growth of primocanes and yield if left undisturbed.

There are two types of damage caused by herbicides, visible and invisible. Visible damage is easy to recognize and can be controlled by careful adherence to label recommendations and careful application. Calibrate your sprayer frequently.

Invisible herbicide damage is much more insidious. It is the result of residual type herbicides. Over time, the plants gradually produce fewer and weaker primocanes. This is sometimes accompanied by minor visible symptoms but they are not severe enough to cause alarm. This type of injury is more likely to occur on light soils or when the same residual herbicide is used year after year.

Reduce the rate of application on light soils. Use only enough to control the weeds. Alternate the type of residual herbicide from one year to the next to prevent a buildup of a single chemical. Be sure you are using the proper rate of application. Overdosage often occurs when making band applications. A two-foot wide band sprayed on the base of the rows of a plantation with 8-foot row spacing is applied at a rate of 100 gallons per acre but since only 1/4 of the area is actually sprayed, it only requires 25 gallons to cover each acre of the plantation.

Nutrition

Nitrogen and water are the nutrients most likely to limit production. If water and other nutrients are sufficient, the amount of nitrogen fertilizer applied regulates plant growth within the limitations of climate and site. Keep track of the amount of nitrogen applied each year and relate it to the amount of growth. Vary it until the desired balance of primocane growth is obtained.

Continue to have the soil tested every three or four years to keep track of nutrient levels. Collect a leaf sample at the same time and have it analyzed for minor elements. Adjust the fertilizer program according to the results. Your agricultural advisor can help you.

Water

The amount of water available during the growing season is very important. It determines primocane vigor and number along with fruit size. Even temporary shortages during critical stages of growth have a marked effect on yield.

An example of this effect is illustrated in Table 9-4. Both the number of vigorous primocanes and yield were low before irrigation was started. During the early years after irrigation was begun, yields were better but not exceptional. Scheduling of irrigation was done during this time by means of soil samples. Only after the shift to evaporation pan scheduling was made in 1965, did the production reach a high level and maintain it from year to year. It seems likely that temporary moisture stresses must have occurred during the irrigation period when scheduling was done by soil sampling.

Pollination

Transfer of pollen from the stamens to the pistils is necessary for drupelet formation (Figure 12-2). This is normally done by wild and tame honeybees. Inadequate pollination results in small crumbly fruit. Usually there are adequate numbers of wild bees in the vi-

FIGURE 12-2. Blackberry flowers. Honeybees gather nectar from the flowers and in the process transfer pollen from the stamens to the pistils of the same flower or others. This transfer of pollen is essential for fruit formation.

cinity to take care of pollination, however, this is not always true, and also, wet cold weather during bloom may reduce their flight. To insure adequate pollination, move one or two hives of bees per acre into the field during blossom time. This insures pollination and provides a good source of excellent honey (Figure 12-3).

Supervision

Carefully supervise all aspects of production, harvesting, and marketing. Careful, well-trained workers are essential for maximum efficiency and profits. Careless, poorly-trained workers cost you money.

Alternate Year Production

In this management system, all canes in half of the field are mowed off at ground level during the winter. The primocanes in that half of the field grow without competition from floricanes during the following summer. There are no floricanes to remove after

FIGURE 12-3. One or two hives of honeybees per acre (4-5/ha) are brought into the field during blossom time to insure pollination during wet, cold, or windy conditions or if wild bees are scarce. (Courtesy of Dr. C. A. Brun)

harvest and the labor required for dormant pruning is minimal. Almost no pesticide applications are needed. These canes are allowed to produce a crop during the following season. In the second year, both floricanes and primocanes are present. The new primocanes are removed or chemically suppressed during the early part of the season. This prevents them from competing with the floricanes during flower and fruit development and results in high yields. Even though suppression or early removal of canes reduces the growth and number of primocanes that year, they have all of the following growing season to recover. The other half of the field is handled the same way in the alternate years.

Research and actual grower experience with trailing blackberries has shown that, although half of the field is out of production every year, the total yield for the two-year period is equal to 80-85% of the total for a comparable field over the same period of time (Sheets, Nelson and Nelson 1975).

This system of management works best for blackberries. Large blackberry growers have used it very successfully to reduce production costs and eliminate much of the cane damage caused by machine harvesting. Red raspberries can be handled in the same manner but the results are not as satisfactory. Both fruit size and quality are reduced. The total yield compared to standard production practices is about 70-80%.

Chapter 13

Harvesting and Marketing

The success or failure of a bramble plantation depends on how successfully the fruit is harvested and marketed. This requires very careful planning and attention to all of the many details involved. By the time the first fruit is ready for harvest, it is too late to make many of the critical marketing decisions that should have been made earlier.

This is not to say that once you decide on a method of harvest or marketing that major changes cannot be made successfully in later years. Growers must study and keep up to date on changes in technology and market opportunities. If, after careful study, a new opportunity appears economically desirable, then the grower should feel free to make the necessary changes. Such shifts in harvest procedures or market outlets are often difficult and expensive. Sometimes it is possible to make them gradually over a period of several years or as old plantings are removed and new ones established. When this can be done, the economical impact is reduced and the changes are less disruptive.

FRESH MARKET SALES

On the Farm

On-the-farm marketing is a business operation and should be approached as such. It requires careful attention to detail and considerable advance planning. In addition to on-farm organization, publicity in advance of the harvest season is essential.

The kind and extent of publicity that is necessary varies with available resources and with the imagination and resourcefulness of the grower. Classified ads in newspapers, leaflets, brochures, direct mail, and personal contacts are among the possibilities. All such notices should include the types of fruit available, telephone numbers, location, business hours, and days of the week when produce is available, along with directions on how to reach the farm. As time goes on, the best and least expensive source of publicity is the word-of-mouth recommendation of satisfied customers. This publicity is earned by providing good facilities, high quality fruit at reasonable prices, and helpful, patient, friendly employees.

U-Pick

This marketing method may be the only one that the grower employs or, as is often the situation, it is used to provide cash flow and to supplement the income from a larger fresh fruit or processed fruit operation. The extent of preparation for handling customers varies with the situation, but the principles are the same.

Large fruited, easy-to-pick cultivars such as the thornless types of blackberries, are most popular. The fruit should be attractive and have a good flavor. Most of it will be used for canning and freezing so it is important to select cultivars that are good for those purposes. New improved cultivars should be considered carefully before one is selected for planting. Farm advisors and neighboring growers who have had experience with the cultivars can be very helpful. Best of all, actual observation of the plants growing in the field, either under commercial or research conditions, is important.

Most growers who produce fruit for on-the-farm marketing prefer to have a long marketing season. By selection of different *Rubus* species and cultivars, it is possible to spread the harvest season from early summer through the first fall frost.

Locate the plantation within easy driving distance of a major population center. Arrange the fields with convenient access to a central control point or provide customers with in-field transportation. Establish convenient parking areas to keep cars out of the fields. In rainy locations, develop a sturdy grass sod or apply a weather-resistant surface to high automobile and foot traffic areas. It is absolutely essential that you provide adequate, clean, well-maintained restrooms and hand-washing facilities.

Keep the fields clean, productive, and well-cared for. Do not use U-pick sales as a salvage operation after the best fruit has been harvested. That is a sure way to develop dissatisfied customers.

Locate the check-in, check-out station close to the parking area. Post a simple, easy-to-understand list of procedures to be followed by the customers in a prominent location. Provide them with plastic pails or other containers to pick the fruit into in the field. Keep a supply of inexpensive containers for them to use in transporting the fruit to their homes. Some growers of thorny blackberries issue a short, 18 in (45 cm) long, wooden lath for the customers to use in parting the bushes so they can reach inside to pick the fruit without getting scratched. A small charge can be made to cover the cost of containers.

Keep a sufficient number of well-trained employees on hand to supervise the customers and keep them happy. Some customers may never have harvested fruit before and will need special instructions on how to pick it without causing damage to the fruit and bushes. Some may, also, need help in transporting the fruit out of the field to the check-out point.

Assign customers to specific rows or parts of rows in such a way that they have adequate, good quality fruit to pick while, at the same time, insuring that all of the fruit that is ready for harvest is picked. Row numbers on the end posts makes it easier to keep track of the rows that have been assigned.

Provide an accurate scale for weighing the harvested fruit. Subtract the empty weight of the container from the total to calculate the selling price. Keep enough change on hand to take care of expected transactions and move all excess cash to a safe place as it accumulates. This reduces temptation and lessens the likelihood of theft.

Some customers prefer to purchase fruit already harvested. Keep a supply in a cool location for this purpose. Maintain an appropriate price differential between U-pick fruit and ready-picked fruit.

People differ greatly in how well they harvest the available ripe fruit. Since fruit becomes overripe and rots if left on the plants, it is a good idea to have a small group of pickers available to go over the fields and harvest the fruit that was missed.

Roadside Stands

Roadside stands vary widely in their complexity. They range from a display table in the shade of a large tree with a single type of fruit for sale, to elaborate buildings complete with refrigerated storage and a wide selection of farm produce. Many successful sales rooms are made by clearing out an existing room in the barn or other building. The produce on display is usually grown on the owner's farm, though some of it may be purchased from other growers for resale. Large operators sometimes enter into contracts with other growers to produce certain crops for them.

Excellent information is available about the construction and operation of roadside markets. This information can be found in various agricultural publications and through local and national farm produce marketing associations. These associations are made up of growers who are actively involved in roadside marketing and are outstanding sources of practical information.

Establish the stand in a location that is convenient and safe for motorists to pull off of the road and to merge with traffic as they leave. The faster and more congested the traffic, the more important it is to develop off-highway, slowdown and reentry lanes.

Many of the customers at roadside stands are impulse buyers, that is, they make up their minds to stop on short notice. For this reason, it is important to locate roadside signs advertising the stand a considerable distance down the road. This gives the potential customers time to make up their minds and start gradually slowing down before arrival at the stand.

Signs should be attractive and easy to read. They should include a description of the types of produce offered together with information on how far ahead the stand is located.

Most potential customers consist of families returning to the city after an outing in the country. For this reason, the best location for a stand is on the traffic-side of the road for returning traffic. Adequate parking must be available and conveniently located close to the stand.

The principal reason consumers purchase from grower-operated stands is to obtain fresh, high-quality, home-grown produce at a reasonable price. Although price is a consideration, freshness and quality are overriding concerns. For that reason, display only the best fruit and maintain it in good condition until sold. Since bramble fruits are especially perishable, make provisions to keep them out of the direct sun and as cool as possible. Keep all that is not needed for display purposes in a cool place or refrigerated room. If adequate holding facilities are not available, have a continual supply of fresh-ly-picked fruit coming to the stand throughout the day. Do not use the stand as an outlet for cull or poor quality fruit unless such fruit is clearly marked as such and sale-priced.

Fresh Market Shipping

Handling bramble fruit for fresh market shipping is a very specialized business and requires careful attention to all aspects of harvesting, handling, packaging, and shipping. The shelf life (the length of time the fruit remains in marketable condition after harvest) varies with the type of fruit. Even with the least perishable, the length of time from harvest to market ranges from only a few days to a week or ten days maximum, depending on how well it is harvested and handled. Red raspberries are the most perishable of the bramble fruits, yet they are the most popular for fresh market shipment. They are followed in order from most to least perishable by raspberry-blackberry hybrids, blackberries, purple raspberries, and finally black raspberries.

Cultivar Selection

Select cultivars that are best adapted for shipment. The characteristics that make them desirable are different from those required for local or processing markets and considerable progress is being made by plant breeders towards development of cultivars for this market (Daubeny 1986; Moore 1984).

Appearance. The best cultivars produce large fruit with many small drupelets. Red raspberries should not darken appreciably during the marketing process. They should be bright red when harvested and retain this bright color during marketing. Blackberries and their hybrids develop a glossy sheen when ripe and the best cultivars retain this glossy appearance at the market.

Firmness. Red raspberries are normally soft and easily damaged during harvesting and handling. This makes them susceptible to infection and breakdown by fruit rot organisms. Some cultivars, especially those with black raspberry in their genetic makeup, are firmer and better able to resist such damage and thus are more resistant to fruit rot. The hollow center allows them to crush during handling and shipping and to settle in the containers. Firm fruit resists this crushing.

Blackberries and their hybrids are also easily damaged during handling. Although they do not crush and settle like raspberries, their skin is very thin and easily punctured, making them susceptible to fruit rot. To date, plant breeders have had only limited success in developing firm, thick-skinned cultivars.

Fruit Rot Susceptibility. There has been some success in breeding fruit rot resistant cultivars. This has been largely due to the development of firm fruit which are less likely to be damaged during handling and thus are less subject to fruit rot infection.

Ease of Detachment. Fruit for shipment should be harvested immediately after it reaches full color. At this stage of maturity, some cultivars do not release easily from the receptacle. This makes them difficult to harvest and more subject to damage because of the force necessary to remove them from the plant.

Harvesting Fruit for Fresh Market

Fruit intended for fresh market must be harvested at the optimum stage of maturity since over ripe fruit has a very short shelf life and is subject to fruit rot. Harvest red and purple raspberries, blackberries, and raspberry-blackberry hybrids every two to three days. Black raspberries remain in good condition on the plant longer, therefore, the harvest interval for them can be extended to five or six days.

More fruit is damaged during picking than at any other stage in the marketing process. Red raspberries and blackberries destined for the fresh market must be picked and handled very carefully. Train your pickers to harvest the fruit directly into the containers in which it will be marketed. This requires careful instruction and supervision. Pick raspberries just as soon as they reach full color. Blackberries and blackberry-raspberry hybrids should have a glossy sheen and separate readily from the stem when tipped to one side. At this stage, the fruit is still firm and has maximum shelf life and optimum flavor when it reaches the market. Drop any overripe, damaged, or malformed fruit to the ground or have it picked into separate containers for local use. If there is much fruit rot in the field, send pickers through the field ahead of time to remove the rotted fruit. This eliminates a source of disease contamination.

Losses caused by fruit rot during transportation and marketing can be excessive unless special care is taken to prevent contamination. Provide clean containers and flats and keep them clean. Use special stands or carts in the field to keep the containers off of the ground until they can be stacked on clean pallets or loaded directly onto trucks.

Postharvest Handling

Once the fruit is picked, it is very important to keep it as cool as possible until it reaches the market. The fruit is alive. It continues to carry on respiration (the intake of oxygen and giving off of carbon dioxide and heat which are the by-products of oxidation of sugars and other materials within the fruit). For each 18°F (10°C) increase in temperature the rate of respiration doubles and this rate of respiration is directly related to the speed with which the fruit softens and breaks down after harvest. Low temperature also slows the rate of infection and development of fruit rot organisms. Since this is so, the best way to slow deterioration of the fruit after harvest is to reduce the temperature.

Harvest in the early morning while the temperature is cool. Keep the fruit out of the direct sun after it is picked and move it into cold storage as soon as possible. Palletize the fruit in the field to speed up handling and reduce the number of times the flats are handled.

The sooner the temperature of the fruit is lowered to near 32°F (0°), the longer its shelf life. The most efficient way to accomplish this is by forced air cooling (Debney et al. 1980). The containers of fruit are exposed to a strong blast of cold air which sweeps through the containers carrying away the heat. An efficient system will reduce the fruit temperature to 35°F (2°C) within two hours. Remove the cooled fruit from the forced air unit as soon as it reaches this temperature since prolonged exposure to rapid air movement dries it out and renders it unmarketable. Once cooled, hold it in regular cold storage at 32°F (0°C) until ready for shipment. The goal of a successful operation is to have the fruit cooled, loaded, and on its way to market within 24 hours after it is harvested. Even at 32°F (0°C) the fruit loses moisture and potential shelf life.

Preparing the Fruit for Shipment

Most of the fruit destined for shipment to distant markets is picked into half-pint (225 ml) containers. This provides a maximum of four layers of fruit and reduces the amount of settling that occurs during shipment. It also decreases the danger of fruit rot development in the lower layers of the container.

The best containers are made of semirigid, clear plastic. They have narrow slots on all sides to provide good air movement through the product during forced air cooling. Rigid plastic provides good protection and allows the customer to inspect the fruit from all sides at the time of purchase. Some containers have hinged lids while others are open on top. The latter are either shipped with the top open or are covered with a thin layer of plastic sheeting to protect the fruit from dehydration during shipment and while it is displayed at the retail market.

Molded paper pulp containers are also used. They are inexpensive, lightweight, can be made of colored material to make them more attractive and are made from a renewable natural resource. Their principle faults are that they stain easily and do not allow the customers to inspect the fruit in the bottom of the container at the time of purchase.

The half-pint (225 ml) containers are packed in corrugated cardboard flats or master containers for shipment and handling. These flats and master containers allow easy stacking on pallets and prevent damage during shipping and handling. They provide attractive space for identification of the grower together with other information. This is especially important in international trade since additional information is often required by customs officials.

Once the containers are stacked on the pallets, stabilize them with strapping to prevent shifting during shipment. High value, long distance shipments are often overwrapped with an air-tight polyethylene film to protect the fruit from dehydration. This air-tight film also makes it possible to inject special gas mixtures into the stack. These mixtures contain high levels of CO_2 which slows the rate of ripening and reduces fruit rot development. The high CO_2 levels dissipate during shipment and have no deleterious effect on fruit flavor. Remove the film wrap or puncture it when the shipment reaches the market, otherwise CO_2 will build up to dangerous levels as the temperature rises in the container. Also, the rising temperature causes the moisture inside of the containers to condense and promotes the development of fruit rot.

Shipment to Market

The fruit can be shipped to market either by refrigerated truck or by airplane depending on the distance. Those markets located less than 1,500 miles away are usually serviced by refrigerated trucks. This method of transportation is cheaper than air shipment and is quite satisfactory for the shorter distances. The temperature of the fruit and the interior of the truck should be close to 32°F (0°C) at the time of loading and should be maintained as near that temperature as possible until it reaches the market. Properly handled truck shipments reach the marketplace in good condition.

Cargoes destined for more distant markets are shipped by air. This method of shipment is expensive but very rapid. The fruit can be delivered to very distant markets within 24 hours. It requires careful coordination of harvesting and handling since the airlines run on a tight schedule and the fruit must arrive at the airport ready for shipment at a specific time.

Market Coordination

The grower should educate himself or herself about all aspects of the fresh market channels before venturing into fresh market shipping. If it is possible to visit the market in the early stages of planning to personally observe handling and marketing procedures, this can be very helpful since any unforeseen delay on the way to market can be disastrous.

Arrange for the sale of the fruit before it is shipped. This can be done directly with the retailer or through agents or wholesalers.

Select personnel that are reliable and familiar with shipping procedures. The worst thing that can happen is for the fruit to sit forgotten and unrefrigerated on the loading or unloading dock. Fruit loss at this point is usually at the expense of the shipper since the business transaction is not complete until the product is delivered in good condition and the money is paid.

Remain in close contact with whoever holds the contract for shipping and handling the fruit. Efficient movement of the fruit with a minimum number of transfers and delays is necessary for profitability. Personnel and schedules can change, often on short notice. Weather and market conditions also vary throughout the season and from year to year. Both the grower and his agent or retailer need to know immediately of such changes so proper adjustments can be made.

Keep a close watch on changing transportation and market conditions so that plans for shipping and marketing can be altered on short notice if the situation requires it.

PROCESSING MARKET

The processing market, while not yielding a high price per unit when compared to fresh market, depends on quantity as well as quality and can be quite profitable and satisfying. Many growers prefer to sell to this market because they do not want to become heavily involved in the attention to detail required for producing fruit for fresh market. If they wish, they also have the option of using machine harvesters, thus eliminating many of the problems associated with recruiting, supervising, liability, housing of workers, and the paperwork involved.

Fruit for processing is utilized in several different ways. Much of it is frozen in bulk containers for institutional use or to be reprocessed into jams, jellies, preserves, pie fillings, yogurt, etc. Some is combined with sugar and used to fill small, retail packages. Good quality, whole fruit is often used for IQF processing. IQF stands for "individually quick frozen." In this process, the whole fruit is spread out in a thin layer on trays and subjected to -32°F (0°C) or lower temperature. It is then packaged without sugar. Overripe, poor quality fruit is often packed in large barrels or drums to be used for juice or wine. A relatively small proportion of the fruit is processed by canning.

Most fruit is sold to processing companies. They set the price and determine how the fruit is to be processed. They also hire field representatives who contact growers before harvest and sign contracts for the purchase of the crop. These agents are usually available for advice on all aspects of culture from the time of planting through harvest. This method of marketing relieves the grower of all responsibility for finding a market for the fruit.

The principal drawback to the processing market is that the grower has no control over the price he receives. This sometimes leads to dissatisfaction, especially during years of high yields and low prices. For this reason, some growers have set up on-the-farm freezing plants and pack their own fruit in 30-pound (13.6 kg) containers. In so doing, they assume the responsibility for finding a market and their success depends on how well they succeed. If successful, they retain the profits that normally go to the processor and at the same time have some control over prices.

Harvesting

Harvest costs for the processing market are considerably less than those for fresh market. The fruit is handled in bulk and, because a wider range of maturity can be tolerated, supervision of the harvest is less restrictive. Pickers can harvest the fruit easier and faster and be paid a lower price per unit. The cost of containers is eliminated, since the processor usually furnishes the flats for transporting the fruit to the plant (Figures 13-1, 13-2). It is also possible to machine harvest.

Handpicking

The care with which the fruit for processing is handled depends largely on the market. More care is required for IQF and small packages, less for bulk packages, and least for barrel packs. The cost of supervision and harvesting is reduced from one type of market to the next and the condition of the product suffers proportionately.

Harvest red raspberries, blackberries, and raspberry-blackberry hybrids every five or six days, more often during wet weather. Space the pickings of black raspberries about a week apart.

Machine Harvesting

Raspberry and blackberry fruits loosen as they reach maturity and will fall off when the plant is shaken. This is the principle upon which machine harvesters operate. Since not all cultivars loosen to the same degree when ripe, some are suitable for machine harvest and some are not.

The machines shake the fruit off of the plant, catch it with special catching plates, and transport it through a blast of air which removes leaves, insects, and other debris. It then travels across a sorting belt where additional insects, rotted fruit, immature fruit, and other debris is removed. The sorted fruit then goes into the flats or other containers ready for transport to market.

FIGURE 13-2. Palletized flats of fruit loaded on a truck and tied down ready for transport.

The efficiency at which the machines operate when compared with handpicking depends greatly on field conditions and the skill of the operator. Under the best of conditions, efficiency is generally considered to be about 80-85%. Most growers are quite willing to suffer this loss in production since machine harvest eliminates most of the problems associated with handpicking. In addition, machine harvesting is suitable for large acreages and does not depend on locating the plantings close to population centers or the establishment of labor camps to guarantee an adequate source of labor. There is very little difference between the two methods in the actual cost per unit of harvested fruit.

Probably the biggest controversy over machine harvest has been over the quality of the fruit harvested. Many processors believe that the fruit is not suitable for the higher priced IQF and consumer package markets. As growers have become more efficient in their operation and the processors have had more experience with machine harvested fruit, this controversy has tended to disappear. Improved harvesters and new cultivars especially adapted for machine harvest are being developed that still further reduce this problem.

The first machine harvesters were built by raspberry growers in the state of Washington (Schwartze and Myhre 1952; Figure 13-3). Even though they resembled present day machines in many ways, none of them were successful and further development did not take place until the late 1960s. The first commercially successful harvester was developed by a family of growers in Oregon named Weygandt to harvest black raspberries. It utilized hinged plywood panels to shake the bushes back and forth. Later they replaced the panels with two large drums, one on either side of the row. These drums had many slender rods that entered into the plants and shook them. The drums were free to rotate as the machine moved down the row. It was suitable for 'Thornless Evergreen' blackberries and red raspberries. Many other machines were developed about the same time by individual growers and some commercial companies with very limited success.

The first widely used machines were developed by Littau in Oregon and by Blueberry Equipment Company (BEI) in Michigan. These machines were patterned after the experimental blackberry harvester developed by the University of Arkansas (Nelson et al. 1965). The shaking mechanism consisted of banks of flexible rods which were pivoted at the front end and could be adjusted for closeness to the row as well as speed and length of stroke. The rods imparted a horizontal shaking motion that caused minimal damage to the plants and was quite effective for fruit removal. These machines were instrumental in shifting a large percentage of the processed red raspberry and blackberry industry of the Pacific Northwest region of the United States from hand harvest to machine harvest.

In recent years, Korvan Industries from Washington State entered the market. Both this company and Littau now manufacture machines which utilize many-fingered vertical drums to shake the plants (Figures 13-4, 13-5). This change has improved fruit removal efficiency but there are still too many fruits lost on the ground in the center of the plants. Improvement in catching mechanisms and cultural methods can be expected to reduce this problem. At the

FIGURE 13-3. Machine harvester built by a raspberry grower in the Puyallup Valley of the state of Washington around 1950. It was not commercially successful. Machine harvesting did not become a commercial reality until the late 1960s. (Courtesy of Dr. C. D. Schwartze)

FIGURE 13-4. Littau machine harvester equipped with rotary shakers for fruit removal.

present time, about 90% of the processing acreage in northwestern United States and British Columbia is machine harvested.

The shift from hand harvest to machine harvest presented an entirely new set of problems. One important result was psychological. Growers had to accept the fact that the plants look much different after the machine has gone over the field than after they have been handpicked. Many of the less ripe fruit that handpickers remove are left on the plant. The normal grower's response was to increase the speed and vigor of the shaking action to remove more fruit. This resulted in more green fruit and fruit with stems still attached, caused more plant damage, and increased the amount of leaves and trash that had to be removed.

Insects that normally remain in the foliage during hand harvest are shaken loose, fall into the fruit, and must be removed. They require extra insect sprays before harvest to remove them.

Machine harvested fruit is more mature and softer, increasing its susceptibility to fruit rot both during and after harvest. This fruit rot

FIGURE 13-5. Littau harvester operating in the field.

(*Botrytis*) is a major problem. Once the rotted fruit is mixed in with sound fruit, it is nearly impossible to remove it. Only a limited amount can be removed by hand on the sorting belt. A spray program to control *Botrytis* is only partially successful, since rainy weather during harvest nearly always causes an outbreak of the disease even when sprays for control have been applied.

The problem can be decreased by shortening the interval between harvests, slowing the ground speed of the machine and, by increasing the number of people at the sorting belt. When these measures are not enough, some growers run the machine through the field and drop the rotted fruit on the ground until dry weather returns and the fruit rot levels diminish to manageable levels (Figure 13-6).

Operation and Adjustment of the Machine. The harvester is expensive and very complicated. It requires careful maintenance and

FIGURE 13-6. Off-loading fruit directly from the harvester into a refrigerated trailer to prevent fruit rot and maintain the fruit in good condition until it can be processed.

constant readjustment during harvest to adapt it to changing temperature, moisture, cultivar, and crop conditions.

The appearance of the harvested fruit when it reaches the sorting belt is a good indication of how well the machine is adjusted. If there are too many green fruits or fruits with stems still attached, reduce the vigor of the picking action. Increase the speed of the air cleaner during wet weather or when excessive numbers of insects or leafy trash appear. Be careful when making this adjustment since too much air will blow ripe fruit out of the discharge. With heavy fruit production, there may be too much fruit passing over the belt for the hand sorters to keep up. Reduce the ground speed of the machine to where they can properly inspect the fruit or increase the number of sorters.

It should be apparent by now that the responsibility of the grower is not finished when the machine harvesters are in the field. Careful supervision is still necessary. It is especially important that either the grower or a reliable employee continually monitor the picking operation. This means following the machine through the field from time to time to check on fruit removal, plant damage, and fruit on the ground. It also means checking the harvested fruit and making proper adjustments.

It is better to spot problems and correct them during harvest than to wait until an emergency call comes from the processor or field representative. Problems that are not corrected before the fruit reaches the processor can be quite expensive.

Chapter 14

Insects and Diseases

ARTHROPOD PESTS

The arthropod pests of brambles include both insects which have six legs and mites and spiders which have eight legs. For convenience, they are grouped together. Insect and mite populations usually build up over a period of time and can be controlled with spray applications when they reach dangerous levels. This level is known as the economic threshold, the level that can be tolerated before it becomes necessary to spray. It varies widely, depending on the type of injury and part of the plant infested. Insects or mites which affect the leaves, stems, or roots can be allowed to build up to moderate levels before it is necessary to control them. Control of those which damage the blossoms or fruit, however, must be started at quite low levels of infestation. The time of year when damage occurs also determines threshold levels. In general, damage to the plants during blossom and fruit formation is most serious and, therefore, threshold levels are low during this period.

The problems caused by arthropod pests vary in their effect on bramble production. Some reduce the growth or quantity and quality of the fruit while others have little effect on production but may be serious contaminants of the harvested fruit. Contamination is a much more serious problem when the fruit is machine harvested than it is for hand harvested fruit. Machine harvested plantations, therefore, require a more intensive insect control program than is necessary for hand harvested fields. It is a common practice for those growers to apply one or more broad-spectrum pesticide sprays just before harvest to eliminate as many of the potential contaminants as possible.

The high cost of insecticides, the tendency of some insects to develop resistance to specific chemicals after many repeat applications of the chemical, the danger of environmental and ground water contamination, pesticide residues on the product, and the current effort on the part of both growers and the consuming public to keep the use of chemicals to a minimum, all emphasize the need for growers to monitor the level of arthropod populations and to spray only when economically necessary.

This requires careful examination of the field at frequent intervals for evidence of insect injury. Once injury is noticed, it is then necessary to identify the insect that caused the injury and to keep track of the amount of damage.

Positive identification of the insect allows the grower to select the proper chemical for control. The insects described herein are those which are most widely distributed. For a more complete description of bramble insects together with their life histories and description, consult local agricultural advisors and publications by Ellis et al. (1991), Pritts and Handley (1989), and Gerber (1984).

The chemicals that can be used for control of bramble pests often change from one year to the next. Specific chemical control measures are, therefore, not listed. Once the pest is identified, follow local recommendations to control it.

Root Weevils

Description and Damage

There are several kinds of weevils that attack brambles (Antonelli, Shanks and Fisher (1991). The hard-shelled adults have elongated, snout-like mouth parts and cannot fly. They emerge from the soil and feed on the buds and leaves at night.

The black vine weevil (*Otiorhynchus sulcatus* F.), the strawberry weevil (*O. ovatus* L.) and the obscure root weevil (*Sciopithes obscurus* Horn.) are commonly found in the United States. Clay-colored weevils (*O. singularis* L.) are more common in Great Britain and Europe. They all reduce yields by feeding on the young, developing buds of the fruit laterals during early spring.

Although the larvae (the immature, wingless, C-shaped grubs that hatch from the eggs) of the weevils feed on the roots and crowns, and the adult beetles chew notches along the edges of leaves, the actual damage caused by this feeding is seldom serious. Weevils are a much greater problem as contaminants of the harvested fruit. They become dislodged from the foliage during harvest and fall into the containers of fruit where they must be picked out by hand. This problem is especially serious with machine harvested fruit.

Control

The presence of weevils is most often discovered during harvest. Feeding notches along the edges of the leaves is also a good sign of their presence, however, other insects may cause similar damage. Most of the adults hide in the soil during the day and come out at night to feed on the foliage. Their population can be monitored by spreading a cloth out on the ground after dark and shaking the plants to dislodge the adults.

Some control can be obtained by keeping down the population of perennial weeds in and around the field. Insecticides, however, are the principal means of control.

Crown and Cane Borers

Description and Damage

The raspberry crown borer (*Pennisetia marginata* Harr.) attacks all members of the *Rubus* family. The adult is a clear-winged, black and yellow moth which lays its eggs on the lower leaves in late summer. The larvae hatch in the fall and crawl down to the base of the canes where they overwinter. They feed on the bark of the canes for a brief period of time in late winter before they tunnel into the crown and canes. They continue to feed and enlarge over a two-year period before they pupate and become adults. Since feeding takes place inside of the crown and the basal portions of canes, the damage is often overlooked. Individual primocanes wilt and die during midsummer. A sharp pull on these canes will cause them to break off at the point of damage and reveal the tunnel in the center of the cane. Most often, their presence is discovered when pruning off the dead floricanes after harvest. Holes in the center of the stubs is evidence of their presence.

Control

Apply an insecticide as a drench to the base of the canes during the winter. This kills the young larvae as they feed prior to tunneling into the crown. Once they are inside the crown, they cannot be killed with insecticide applications. Since they have a two-year life-cycle, you must apply an insecticide every year for two or more years to kill all of the newly emerged larvae.

Other Cane Borers

Two other species of borer cause damage that is similar to that of the raspberry crown borer. The red-necked cane borer (*Agrilus rificollis* Fabricius) and the cane maggot (*Pegomya rubivora* Coquillett) tunnel into the aboveground portion of the canes. Chemical control is usually not necessary since most of the larvae can be removed with the dead floricanes when they are pruned off after harvest or by cutting back the injured tips of primocanes when the damage is observed.

Cutworms

Description and Damage

Larvae of the variagated cutworm (*Peridroma saucia* Hubner), the spotted cutworm (*Amathes c-nigrum*) and the double dart moth (*Graphiphora augur* F.) climb the canes at night and feed on the buds and leaves of new growth. Severe infestations can completely defoliate portions of the field. They tend to be concentrated in localized areas within the field and are not present every year. If present during harvest, they fall into the harvested fruit and must be picked out. The larvae should be killed rather than dropped to the ground where they will climb back up the plant and continue to cause damage.

Control

Watch for signs of feeding. When the amount of damage becomes severe enough to warrant control, spray the field with a recommended insecticide. Late evening or nighttime applications when the larvae are actively feeding are the most effective.

Leaf Rollers

Description and Damage

The larvae of both the orange tortrix (*Argyrotaenia citrana* Fernald) and the oblique-banded leafroller (*Choristoneura rosaceana* Harris) conceal themselves within folded young leaves and feed on the developing foliage. Occasionally they feed on the fruit. The amount of foliar damage they cause is minor, but they are serious contaminants of harvested fruit.

Control

Leaf rollers have many insect and spider predators that keep down the populations if left undisturbed. Unfortunately, sprays to control the leaf rollers, also kill the predators. Therefore, it is necessary to monitor the number of leaf roller adults and larvae present during the period close to harvest. Traps baited with pheromones that attract the adult males can be used to assist in the monitoring. Apply sprays for control only when necessary (Knight et al. 1988). One insecticide application within a week of the beginning of harvest will usually provide adequate control.

Spider Mites

Description and Damage

Spider mites have eight legs and resemble tiny spiders about 1/32 in (0.5 mm) long. They range in color from straw to green and red. The most common species is the two-spotted spider mite *Tetranychus urticae* Koch. Other species occur which have similar life cycles and cause similar damage. Almost always, there are predator mites present that feed on the spider mites and help keep them under control. The eggs are laid on the underside of the leaves where the adults hatch and feed by sucking out the sap. This causes a light stippled effect on the upper sides of the leaves where the green chlorophyll has been removed. The leaves turn yellow or brown and the growth of primocanes is stunted. A thin webbing is usually noticeable on the bottom of the leaves but a magnifying glass is necessary to see the adults. Severe infestations cause early leaf drop and increase the danger of winter injury (Doughty, Crandall and Shanks 1972).

Control

Light infestations cause very little damage and do not require control measures. Such populations will most likely be held in check by natural predators. Sprays for control kill the predators and often result in a subsequent rapid and damaging build up of spider mites. For this reason, sprays to control other insects during the early part of the season should be selected that have little or no effect on mite predators.

Mite populations increase rapidly under hot, dry, dusty conditions and often build up during late summer. Apply miticides only if there is excessive discoloration of the leaves. In regions where severe mite damage can be expected every year, apply a dormant spray. This will usually control early season infestations.

Raspberry Bud Mites

Description and Damage

These mites (*Phyllocoptes gracilis* Nalepa), often called dryberry mites because of the type of damage they cause, are microscopic in size and belong to a group of eriophyid mites. They overwinter under bud scales and in crevices of the bark. The number of mites increases rapidly in the spring when they feed on the new foliage causing stunted growth and irregular, yellow patches on the leaves. Individual drupelets of raspberry fruit turn red prematurely causing misshapen fruit. Young blackberry fruit dries up soon after bloom and fails to mature. The damage is sometimes confused with that caused by virus diseases.

Control

Apply systemic miticides or sulfur-containing compounds in the spring before bloom. Apply the sprays at high pressure and volume for best results.

Redberry Mites

Description and Damage

The redberry mite (*Acalitus essigi* Hassan) is an eriophyid mite that attacks blackberries causing the fruit to remain hard and green or turn red without fully ripening. It is most damaging to late maturing cultivars.

Control

Spray with sulfur compounds during the winter and again in the early spring.

Raspberry Fruitworms

Description and Damage

The larvae of three species of beetles (*Byturus bakeri, B. rubi* Say) and (*B. tomentosus* De Geer) are most commonly found on raspberries, though they sometimes attack blackberries. The small, yellowish-white larvae tunnel into the receptacle of the fruit causing it to fall or they may remain in or on the fruit at harvest time. Adult beetles are about 3/16 in (4.5 mm) long and yellow to brown in color. They feed on both flower buds and leaves. They eat elongated holes in the leaves early in the spring causing them to develop a tattered or skeletonized appearance.

Control

Apply sprays during the prebloom period or at the green fruit stage.

Other Insects

Several other insects are sometimes found in bramble plantings. They vary greatly in the severity of infestation and the amount of damage they cause. Some of those found on the foliage are aphids, blackberry psyllids, western winter moths, raspberry sawflies, and leaf midges. Damage to canes may be caused by rose scale, tree crickets, and raspberry cane midges. Those which are found on the flowers and fruit include several miscellaneous beetles and rose chafers. There are also Lygus bugs and raspberry bud moths.

You must be on a constant lookout for the presence of insects and evidence of their feeding. The need for and the method of control that you use depends on the severity of damage and the type of insect. If there is any doubt as to the identity of the insect, collect a specimen and take it to an agricultural advisor for identification and recommendations for control.

FUNGUS AND BACTERIAL DISEASES

Diseases generally cause more serious damage than insects and their control is more difficult. Whereas it is possible to wait until insects have reached economically important levels and then bring them under control, this is not the situation with diseases. Their control almost always depends on prevention of infection. Once infection has taken place, diseases can seldom be eradicated with spray applications. Control measures are, therefore, limited to the prevention of their spread to healthy tissues. This is accomplished by (1) elimination of the source of infection, (2) provision of environmental conditions unfavorable for development of the disease organisms, (3) the application of toxic materials to the plant surface that kills the disease before it has a chance to invade the tissue, and (4) use of cultivars that have natural resistance to the disease.

The number of diseases that attack brambles is far too numerous to include all of them in this publication. The following descriptions list only the more common and most serious ones. For a more complete listing of diseases and a comprehensive description of their life cycles and control, consult Ellis et al. (1991) and Jennings (1988).

Root Rots

Most root rot damage occurs in heavy, wet soils. Much, therefore, can be done to prevent its occurrence by choosing a well-drained, disease-free planting site. This is not always possible nor is it a guarantee that root rots will not occur.

Phytophthora Root Rot

Root rot caused by *Phytophthora* spp. is the most damaging of the root rots. Most red and purple raspberries are seriously damaged by this rot, but some black raspberry cultivars and raspberry-blackberry hybrids are also susceptible. It is more commonly found in heavy, wet soils but is not limited to them. Once infection of susceptible cultivars takes place, it often kills out entire sections of the planting.

Symptoms. The first visible evidence of its presence in a field is wilting and dieback of the terminal portions of new primocanes as temperatures increase and stress conditions develop in early summer. This wilting and dying increases in severity over one or two seasons and eventually the plants die. Infected canes have a dark, water-soaked region at their base. The roots and crown reveal reddish-brown interiors when scraped. Improved soil drainage or decreased soil moisture has little effect on the progression of the disease.

Control

Prevention of soil contamination is the best control. The disease is most often spread from one location to another in infected soil or planting stock. The plants seldom show signs of infection during the first year after the fields have been fumigated for disease control (Johnson, Crandall and Fisher 1972). This gives a false sense of security since planting stock dug from such fields looks normal yet carries the disease and will infect new fields. It is important, therefore, to purchase certified stock from reliable nurseries which have no history of root rot. The use of tissue culture plants also gives added insurance that the stock is free of *Phytophthora* root rot.

There are marked differences in the degree of susceptibility to phytophthora (Barritt, Crandall and Bristow 1978) among red raspberry cultivars. 'Latham' is the most resistant followed in order by 'Newburgh' and 'Sumner'. Other cultivars range in susceptibility from moderate to very severe. On sites where phytophthora is diagnosed or suspected, the selection of more resistant cultivars helps to overcome the problem.

Soil fumigation prior to planting reduces the population of the disease organism and assists in the establishment of new plantings, but the effect is only temporary and after two or three years the plants begin to die. Plants taken from such areas contain the disease and are capable of spreading it to uninfested soils. The organism can also be spread to other fields by the movement of infected soil on tractors and other equipment.

Annual soil applications of a chemical specific for control of phytophthora can be used to maintain normal plant growth in areas where such chemicals are approved for use (Scheer et al. 1993). Such chemical control measures are only one part of an integrated program involving prevention of soil contamination and the use of resistant cultivars.

Wet Soil Root Rot

Heavy, poorly drained soils are the basic cause of this type of root rot. Apparently, roots which are subjected to prolonged periods of waterlogged conditions lose their ability to resist the invasion of weakly pathogenic disease organisms such as *Fusarium, Rhizoctonia,* and *Sclerotinia*. The more severe the flooding and the longer it persists, the more damage these organisms cause.

Symptoms. New primocanes are stunted and yellow. The floricanes reach varying stages of development before they dry up. They often appear nearly normal up until harvest at which time they wilt and die. Floricane symptoms are sometimes confused with winter injury but a combination of standing water during the rainy season, weak primocane growth, and brown or black roots help to differentiate between the two. As the soil dries out and conditions become more favorable for root growth, primocanes often resume near normal growth. The long-term effect, however, is a gradual weakening of the plants and eventually death. Over-irrigation also causes yellowing and weak growth in low, poorly drained sections of the field.

Control. Select a site with medium to light textured soil and good subsoil drainage. Install underground drains in low sections of the field. Small growers and home gardeners can use raised beds. Use cultivars that are resistant to root rot and schedule irrigation carefully to prevent overwatering.

Crown and Root Gall

Symptoms

Galls are caused by bacteria in the soil that invade the plants and cause development of tumorlike masses on the roots, crowns, and canes. They are most often found on the roots and crowns of raspberries and on the canes of blackberries. Infection of new plantings results in poor stands and weak growth. Older plantings may appear normal but growth and production are reduced depending on the severity of the disease. Galls on the roots are hidden by the soil and may go unnoticed (Figure 14-1).

Control

Prevention is the best control. Purchase planting stock that is free of galls and plant it in disease-free soil. Once the plants and soil are infected, there is no good method available for eradication. If you are planting in soils suspected of being infected, dip the plants in a solution of *Agrobacterium radiobacter*, a nonpathogenic, biological control agent that helps prevent infection. Burr et al. (1993) found this material to be ineffective for red raspberries.

Verticillium Wilt

Verticillium wilt is a soil fungus disease that severely damages black raspberries and some cultivars of purple and red raspberries and blackberries. 'Loganberries' are quite resistant but 'Boysenberries' are very susceptible.

FIGURE 14-1. Root galls develop entirely under the ground on red raspberry and may go undetected for some time. They cause poor growth and low production.

Symptoms

The leaves of black and purple raspberry primocanes turn pale green during midsummer but seem to recover during the cool fall weather. The following spring they turn yellow, beginning at the bottom of the canes and then wilt and die. The canes develop a distinct bluish color, often only on one side of the plant, but eventually after two or three years, the entire plant dies.

Red raspberries are not as seriously affected but the leaf symptoms are similar. The plants may continue to live for many years but the canes are stunted and production is low. Primocanes arising at a distance from infected plants grow normally.

Blackberry canes wilt and the leaves turn yellow or brown during late summer. They appear to recover in the fall but the canes that survive the winter leaf out and set fruit and then collapse before the fruit ripens. There is no blue discoloration of the canes. The spring symptoms are often confused with those caused by winter injury or wet soil root rot.

Control

There is no cure for verticillium once infection has occurred. It can be prevented by selecting disease-free plants from a reliable nursery and planting in disease-free soil. Preplant soil fumigation usually provides good control but is very expensive. It is better to avoid soils with a history of verticillium and those in which susceptible crops such as tomatoes, potatoes, peppers, and eggplants have been grown recently. Nonsusceptible crops should be grown for at least four or five years on infected soils before planting with raspberries. Even then, it is doubtful that such soils are safe for black and purple raspberries.

Anthracnose

Anthracnose is a serious cane and leaf spot disease of black raspberries, blackberries, and susceptible cultivars of purple raspberries. It is much less serious on red raspberries.

Symptoms

Severe infection results in weak growth, defoliation, wilting of shoots, increased susceptibility to winter injury, and unmarketable fruit. It causes the most damage to canes and fruit petioles. Small, purple, round, or oval spots develop on the young canes and petioles. These spots enlarge to about 1/8 in (3 mm) in diameter and the centers become sunken and light gray in color while the margins remain purple. These spots often merge with one another forming irregular dead areas that encircle the canes or fruit petioles causing them to die. Less severe damage causes poor growth and weakens the canes making them susceptible to winter injury.

Leaves develop numerous small purple lesions with white centers that sometimes fall out, leaving a shot-hole effect. Flowers and fruit may also be infected. Individual drupelets fail to develop and either shrivel or are slow to ripen. Such fruit injury is especially bad with trailing blackberries. Most of the erect types are quite resistant to fruit infection. Susceptible cultivars of red raspberry develop many small surface lesions on the canes during late summer causing them to have a grayish appearance. These spots remain small and seldom penetrate beyond the surface, therefore, they have little or no effect on growth.

Control

Anthracnose is difficult to control. It thrives under cool, moist conditions and infection takes place from old canes to new canes. Use a combination of sanitization to remove sources of infection, cultural practices that promote good air movement within the plant canopy and rapid drying after rains or irrigation, and apply fungicides to prevent infection from taking place.

Use disease-free planting stock. Cut off the old floricanes close to the ground either after harvest or during the dormant season and thoroughly incorporate the prunings into the soil before growth begins in the spring.

Good air movement through the plants speeds up drying after rainfall or irrigation and reduces infection. Eliminate weed growth in the rows and use a trellis system that separates the new canes from the old canes as much as possible. Maintain the plants in narrow rows and, where anthracnose is severe, reduce the number of canes left after dormant pruning.

Apply a sulfur fungicide just as the buds start to show green in the spring. Timing of this spray is important. It is not only ineffective if delayed too long, but may also injure the new growth. Additional sprays of other fungicides are usually not necessary.

Cane Blight

Cane blight is caused by a fungus that invades the tissue through wounds on the first-year growth of all species of brambles. Once inside, it spreads up and down the new shoots.

Symptoms

There is little evidence of its presence during the summer, but scraping the surface of the cane in the fall will reveal a brown streak extending a considerable distance away from the wound. During the winter, these lesions develop further and become visible as brown or purple areas extending up and down the canes. The canes become brittle and break easily. The lesions often girdle the canes causing them to wilt and die. Lateral buds fail to grow or produce weak, nonproductive laterals. This effect on the laterals and buds resembles that of a similar disease called bud blight. The symptoms are similar except that with bud blight the brown lesions are confined to the base of the buds.

Control

As with anthracnose and other fungus diseases common to brambles, cultural practices designed to allow quick drying of the foliage after rainfall or irrigation help to prevent infection. Since the disease requires open wounds for entry into the tissue, handle the primocanes carefully to prevent damage. Machine harvesters are often a major cause of cane damage, therefore, it is important to combine good training practices with careful adjustment and operation of the machines.

In regions where cane blight is prevalent, summer tip black raspberries and erect blackberries during dry weather so the wounds have a chance to heal before irrigation or rainfall occurs. Apply fungicides before and after harvest or immediately after summer tipping.

Midge Blight

Symptoms

Midge blight is a disease of red raspberries that is primarily confined to Europe. It results from fungal infection of wounds caused by the feeding of the cane midge, a sucking insect that causes small, narrow splits in canes during early spring. Infection of these wounds by the cane blight organism weakens the canes and causes many of them to die during the fall and winter.

Control

Apply insecticides during early spring to control the cane midge. This prevents blight infection by eliminating points of entry.

Rosette

Rosette, or double-blossom, is a serious fungal disease of black-berries in the central and southern parts of the United States. It is commonly found on erect blackberries and causes reduced yields, poor fruit quality, and the death of canes. Trailing blackberries are seldom infected.

Symptoms

Infected plants produce several leafy shoots from each infected lateral bud in the spring. These shoots are short, have smaller than normal pale green or reddish leaves, and give a "witches-broom" appearance. Bloom is delayed and the flowers become wrinkled, twisted, and pinkish and resemble double blossoms. Infected flowers either fail to set fruit or produce small, abnormal fruit.

Control

Plant cultivars that are resistant to rosette in regions where it is difficult to control and use disease-free plants. Eliminate all wild blackberries growing in the vicinity. Cut out infected canes. In more seriously infected fields, mow off and destroy all canes right after harvest. Several applications of fungicide at weekly intervals during blossom time combined with good sanitation practices will usually give satisfactory control.

Powdery Mildew

Powdery mildew develops during warm, dry weather. It is more commonly found on raspberries but does occur on blackberries as well. Susceptibility to infection varies widely among cultivars.

Symptoms

The upper surface of infected leaves develops a blotchy yellow appearance and the underside below these yellow blotches is covered with a white powdery material. Infection of the tips of new growth causes them to grow long and spindly with small curled leaves. The powdery white fungus attacks the fruit rendering it unmarketable.

Control

The most satisfactory means of control is to plant resistant cultivars. Mild infections do not cause sufficient damage to make chemical control necessary. Removal of infected shoots during late summer helps to reduce infection. Two chemical sprays applied during early bloom usually give satisfactory control.

Downy Mildew

Downy mildew is favored by cool, damp weather. It is especially serious on blackberries and their hybrids in which it becomes systemic and spreads throughout the entire plant. It overwinters in the roots, crowns, and canes.

Symptoms

Under favorable temperature and humidity conditions in the spring, it attacks all of the aboveground parts of the plant. Infected flowers and fruit dry up and the upper surface of the leaves develops yellow, red, or purple angular spots. These spots are accompanied by white or gray lesions on the lower surface. As the leaves age, the spots develop yellow or brown margins and the leaves fall prematurely. New primocanes are stunted with reddish terminal leaves. The fruit fails to develop a typical glossy appearance and dries up without ripening.

Control

Purchase plants from a reliable nursery. Maintain good air movement within the plant canopy and apply systemic fungicides during the early spring to protect the new foliage. Keep the number of sprays to a minimum and alternate types of fungicides frequently. This is necessary to prevent the mildew organism from developing resistance to the chemicals.

Orange Rust

Orange rust attacks erect blackberries, most trailing blackberries, and both black and purple raspberries. Most red raspberries are resistant. It does not kill the plants, but they are greatly weakened and yields are reduced.

Symptoms

Symptoms develop as soon as growth begins in the spring. Many weak, spindly shoots with small, pale green or yellow leaves are produced. These leaves soon produce blisterlike bumps on their undersides. The blisters rupture and produce a mass of orange-colored powdery spores. Infected leaves drop and new, normal-appearing leaves develop. The disease overwinters in dead leaves on the ground and within the plant tissues.

Control

There are no fungicides known to give effective control of this disease, though current research with new chemicals is promising and one may be discovered soon. There are some resistant cultivars. In areas where the disease is expected, these can be utilized. Red raspberries are resistant but all black and purple raspberries are susceptible. Other than the use of resistance, eradication of sources of infection is the best control. Early spring is the time when infection occurs and when it is easiest to recognize. Remove diseased plants and infected shoots immediately along with all infected wild brambles surrounding the field.

Gray Mold Fruit Rot

Gray mold caused by *Botrytis cinerea* is the most widespread and most costly disease that attacks brambles. Red raspberries are more susceptible than black raspberries and blackberries but it can greatly reduce the post harvest shelf life of all of them.

Symptoms

The spores that cause infection are nearly always present and when moisture and temperature conditions are right, will infect ripe fruit and blossoms. Infected fruit deteriorates rapidly becoming watery and soft. Tiny, black or gray spores develop on the surface of the fruit which, under favorable conditions, infect adjacent fruit.

Development of the disease is favored by moisture and high humidity. Prolonged rainy periods during harvest are potentially disastrous and require careful harvest procedures and fungicide sprays to keep fruit rot under control. Fruit harvested under such conditions is not suitable for shipment to fresh market. Even with refrigeration, the shelf life of the fruit is very limited. Sprinkler irrigation does not usually have much effect on infection since it is most often applied during good drying weather and the foliage dries off before infection can take place. However, if the application coincides with a period of cloudy, wet weather it can result in serious damage.

Control

Control consists of an integrated program involving special cultural and handling procedures combined, in some situations, with fungicidal sprays. There is some genetic resistance to infection correlated with fruit firmness (Jennings 1988). Use training systems and cultural practices that open up the plant canopy and provide good air movement. Delay irrigation until two or three days of dry, sunny weather is forecast for the period after application.

Since ripe and overripe and damaged fruit is very subject to infection, decrease the interval between pickings and handle the fruit very carefully to prevent damage. Remove it from the field as soon as possible and either place it in cold storage or process it immediately. Keep the time in cold storage as short as possible since the cold only slows the rate of growth but does not stop it.

Harvest fruit destined for shipment to fresh market at an early stage of maturity and place it directly into the container in which it is to be marketed. Cool it to near 32°F (0°C) as soon as possible and maintain it at that temperature until sold to the consumer. Protective sprays help to reduce infection, but cannot be relied upon to give complete control. Apply them at ten to 14-day intervals beginning at full bloom and continued into harvest. In regions where climatic conditions are warm and dry during this period, omit the sprays. If, however, wet weather occurs, an outbreak of infection can be expected and preventive sprays should be applied.

PHYSIOLOGICAL DISEASES

Crumbly Berry

Crumbly berry is a catchall term used to describe a situation where not enough drupelets develop to form a normal fruit. The drupelets are only loosely joined and when the fruit is picked, it crumbles. Any factor that prevents adequate drupelet set can cause it. Crumble occurs in red raspberries and may be caused by, but is not limited to, infection by tomato ringspot virus or raspberry bushy dwarf virus. The latter is a pollen-transmitted virus that causes severe crumble in some cultivars (Daubeny, Stace-Smith and Freeman 1978).

Occasionally a cultivar that normally produces good fruit mutates to a strain that has poor pollen (Daubeny, Crandall and Eaton 1967). Usually this only affects a single plant, however if it occurs during plant propagation, all plants produced from the mutated plant are crumbly and the results can be disastrous (Figure 14-2).

Crumble can also result from any factor that prevents pollination or reduces the viability of the pollen. Among those which have been noted are: (1) damage to the flower parts by *Botrytis* fruit rot, (2) frost damage, (3) wet soil root rot, (4) boron deficiency, (5) insect or mite injury to flower parts, and (6) inadequate transfer of pollen by insects. The latter can be caused by insufficient numbers of bees in the vicinity of the field, though, it is often the result of careless application of insecticides. Some insecticides kill large numbers of bees in the field and others are carried back to the hive where they kill additional bees. Other spray materials may inhibit bee activity within the field for a week to ten days after application. Even though brambles are generally self-fruitful, they need insect activity to move the pollen from the anthers to the pistils within the individual flowers (Shanks 1969).

Control

Control of crumble, therefore, depends upon its cause. Once this is determined, appropriate control measures can be taken. A simple

FIGURE 14-2. Crumbly 'Sumner' fruit caused by a genetic mutation. Undetected during propagation, the mutated plants were distributed widely.

shift in cultural practice may be all that is necessary. In more complicated situations, it may be necessary to remove the field and start over again using proper precautions to prevent a reoccurrence.

VIRUS DISEASES

Numerous virus diseases infect brambles. Some of them are quite serious. For a comprehensive review and good diagnostic pictures refer, to Converse (1991). They are grouped according to the method of transmission as follows: (1) aphid-transmitted, (2) pollen-transmitted, and (3) nematode-transmitted. All of them can be transmitted from one plant to another by grafting.

Viruses are systemic, that is, they travel throughout the entire plant, so any new plant that is developed by vegetative propagation such as suckers, tip layers, leaf cuttings, or root cuttings is infected.

Even tissue culture plants are infected unless special care is taken to rid the tissue of virus before it is multiplied. Black and purple raspberries are more seriously damaged by viruses than either red raspberries or blackberries. There are also big differences in susceptibility among individual cultivars.

Control

Once a plant in the field is infected, it is not possible to eliminate the virus. The only method of control is to isolate healthy plants from sources of infection or to eliminate the vector that transmits it from one plant to another. This is why it is very important to insist on virus-free planting stock when establishing a new field. Since black raspberries are so susceptible to damage and many red raspberry cultivars can thrive with few or no visible symptoms, it is generally recommended that new black raspberry plantings be separated from old red raspberries by at least 600 ft (180 m) to slow transmission of the viruses from one to the other. In commercial districts where only virus-free planting stock is used, black and red raspberries are sometimes grown side-by-side with few virus problems. In recent years, new red raspberry cultivars have been developed that are resistant to feeding by the aphid that transmits viruses from one plant to another (Daubeny and Stary 1982). The utilization of such cultivars eliminates the transmission of viruses to or from them. Nematode-transmitted viruses can be prevented by elimination of nematodes by soil fumigation prior to planting.

Only a few of the many viruses that infect brambles are widely distributed and cause serious damage. The others are either quite limited in their distribution or do not cause serious economic loss.

Raspberry Mosaic Virus

Raspberry mosaic is an aphid-transmitted virus that causes serious damage to black and purple raspberries. It is less damaging to red raspberries and practically harmless in blackberries. Aphids feed on diseased plants and carry the virus to healthy plants. Even though a diseased plant may appear normal, it still carries the virus. The disease can spread throughout an entire field in two or three years (Figure 14-3).

FIGURE 14-3. The weak 'Newburg' raspberry plants in the center of the row are infected with mosaic virus. (Courtesy of Dr. H. A. Daubeny)

Symptoms

The leaves have light green or yellow blotchy areas. In the more advanced stages, these areas develop large green blisters and the leaves are small and sometimes deformed. The plants of susceptible cultivars become stunted and eventually die.

Control

Use virus-free planting stock, select virus-tolerant cultivars or those upon which the aphid vector does not feed, and isolate the planting from sources of infection. Locate new plantings of black and purple raspberries at least 600 ft (180 m) from older plantings and wild species. Spray programs to control aphids help to slow the spread within the field but cannot be depended upon for complete control.

Raspberry Leaf Curl

Leaf curl is more commonly found in red raspberries but may also infect both black raspberries and blackberries. It is transmitted by an aphid species that is sluggish in its movements and spreads the virus within the field slowly.

Symptoms

Symptoms appear during the season following infection. Leaves of both new and old canes of red raspberries turn light yellow. On black raspberries, they become dark, greasy green. The leaves are small and severely curled downward at the tips and around the edges. Growth is stunted and the fruit is small and of poor quality. The plants slowly become more and more stunted and weak.

Control

Use certified virus-free plants and isolate plantings from infected fields and wild brambles. Since the virus spreads so slowly through the field, it is possible to maintain satisfactory control by digging out infected plants as soon as they are detected. Eventually the number of missing plants will render the field uneconomical for further production. This level of missing plants is usually considered to be about 15-20%.

Tomato Ringspot Virus

Tomato ringspot is transmitted by dagger nematodes (*Xiphinema* spp). It is most serious in red raspberries but occasionally occurs in blackberries and their hybrids. Even in the red raspberry, some cultivars can be infected and show little or no visible damage. The virus is often present in common weed species and if the dagger nematode is present in the field, they will transmit it from the weeds to the raspberries.

Symptoms

Infected cultivars become weak and may eventually die. The fruit is small and crumbly. The pattern of spread within the field is spotty and develops in a target-area pattern that is typical of a nematode infestation. Some of the shaded leaves on newly infected plants develop a ringspot symptom during early summer but these ring-spots soon disappear and do not reoccur on the same plants.

Control

Plant ringspot-resistant cultivars using virus-free planting stock. Test the soil before planting and if dangerous nematode levels are found, fumigate the soil.

Chapter 15

Pest Control Procedures

The soil and air environment around bramble plantations contains numerous organisms such as wild animals, plants, birds, fish, bacteria, fungi, and insects. Most of these live in harmony with the brambles and cause little or no damage. Sometimes they are actually beneficial. Bees and other flying insects are essential for pollination and must be protected from pesticides. Other beneficial insects and/or diseases act as natural predators of harmful insects. By providing conditions favorable for their development, it is often possible to maintain pest damage within acceptable levels without pesticides.

ARTHROPOD PEST CONTROL

Most of the materials used for arthropod control are not specific. They kill beneficial organisms as well as the targeted pests and make a repeat buildup of the harmful pests likely. In addition, repeated use of a chemical may cause the pests to build up resistance and make the pesticide useless.

When alternative methods fail to give adequate control, chemicals are useful. Select the best one based on an accurate diagnosis of the problem. Even then, you must decide whether it is necessary to spray. Carefully evaluate the amount of damage present and the potential control that can be expected from beneficial predators and parasites. Apply pesticides only after the damage reaches unacceptable levels.

DISEASE CONTROL

Disease infections are most easily controlled with preventive sprays applied during the early stages of infection. Since weather conditions have a major influence on the development of many diseases, a change in the weather pattern may determine whether a spray is necessary. However, weather is not always predictable and applications for disease control must often be predicated upon past experience with the infection.

PESTICIDE APPLICATION

Current emphasis on sustainable agriculture, the need for minimal pesticide residues on the fruit, prevention of environmental pollution, and the high cost of pesticides all present a real challenge to bramble growers to use as few and as little pesticide as possible and yet maintain the health of the planting and satisfactory fruit quality. Fortunately, in spite of the large list of diseases and insects that may attack brambles, many handpicked plantations go for years with few or no pesticide applications. Proper site selection, use of the best adapted cultivars and virus-free plants, good sanitation practices, and careful monitoring of fields for signs of disease infection and levels of insect populations all enter into the number of pesticide applications necessary. If all precautions fail and infestations occur, then it is extremely important that the right pesticides are applied in the proper amounts and the most effective manner.

Choice of Materials

Once the problem is positively identified, a pesticide can be selected to control it. Government regulations restrict the materials available for use on brambles and it is illegal to use any that are not listed for the crop or for the purpose for which it is to be used. Even then, there are usually more than one from which to choose. The choice depends upon the stage of plant development, its proven effectiveness, and the effect on the environment, including its effect on beneficial predators and pollinating insects.

Wild and domestic bees are essential for high yields and top fruit quality, hence when sprays must be applied during blossom time, it is very important to use a chemical that will cause the least damage to them. When it is necessary to use a harmful pesticide, the severity of injury can be reduced by applying the spray during late evening after the bees have returned to their hives.

Application

Several different types of sprayers are used for brambles. Those for soil application range from one or two nozzles on each side of the tractor to spray narrow bands of herbicide on the soil at the base of the plants, additional nozzles across the back to include the row middles, or boom sprayers up to 20 ft (6.1 m) or more wide for preplant applications. These machines are designed to apply the spray material uniformly and accurately to the soil surface. The rate of application must be measured and checked frequently to insure accuracy since the nozzle openings wear rapidly from the abrasive nature of some herbicides. Calibration of the sprayers for accuracy is extremely important and relatively simple (Table 9-1).

Maintain spray equipment in good condition and adjust it to deliver the correct amount of material at the correct pressure. Applications at greater than recommended rates are costly and may result in illegal residues in or on the fruit.

Weed Control Sprayers

Herbicides are usually applied with some type of boom sprayer (Daum and Reed 1983). They apply the material at relatively low pressure and with great accuracy. The rate of application per unit area for herbicides is especially critical because too much can cause crop damage and too little will result in inadequate weed control.

Insect and Disease Control Sprayers

Pest control sprays must be broken up into very fine droplets and deposited uniformly on all of the foliage. This requires either a large amount of water at high pressure or a small amount of water and a huge volume of rapidly moving air to break the water up into fine particles and drive it throughout the foliage. The types of sprayers used to accomplish this varies with the size of the acreage and the wishes of the grower.

Large water volume, high pressure, conventional sprayers with various arrangements of nozzles to distribute the material are commonly used for small- to medium-sized plantings (Ross 1987). These apply from 100 to 300 gal per acre (154-454 1/ha) depending upon the density of the foliage. In areas near human habitation and under windy conditions the drift from such applications is apt to create environmental problems. A hooded spray boom can be used to reduce the drift and increase the efficiency of the application (Shanks, George and Crandall 1972).

Larger growers often use various types of air blast sprayers (Pritts and Handley 1989) because they require less water and are more efficient for covering large acreages. They have big fans which deliver large volumes of air at a high velocity. This breaks up the spray material into small droplets and drives it through the foliage (Figure 15-1). Drift from such applications can be excessive

FIGURE 15-1. Air-blast sprayers use high-volume, rapidly moving air to break the spray material into tiny droplets and drive it throughout the foliage.

under windy conditions but with large acreages this is seldom a problem.

PESTICIDE SAFETY

Many of the materials available for commercial pest control are toxic to humans and other animals. It is always best to use low toxicity chemicals for control but this is not always possible and more hazardous materials must be employed. Special safety procedures must, therefore, be taken to protect the personnel involved in the application as well as those who may be present during or after the application.

Learn about the precautions necessary for safe use of the pesticides that you use and include them in your spray program. If such safety precautions are followed faithfully, pesticides can be used safely.

Bibliography

Achmet, S., L. Kollányi, A. Porpáczy and K. Szilagyi. 1980. Procedures for the production of virus-free stocks of small fruits in Hungary. *Acta Hort.* 95:83-85.

Antonelli, A. N., C. H. Shanks, Jr. and G. C. Fisher. 1991. Small fruit pests. Biology, diagnosis and management. Wash. Coop. Ext. Serv. EB 1388.

Bailey, L. H. 1941-1945. Species Batorum. The genus *Rubus* in North America. *Gentes Herb.* 5:1-918.

Barritt, B. H., P. C. Crandall and P. R. Bristow. 1978. Breeding for root rot resistance in red raspberry. *J. Amer. Soc. Hort. Sci.* 104:92-94.

Barritt, B. H., P. C. Crandall and P. R. Bristow. 1981. Red raspberry clones resistant to root rot. *Fruit Var. J.* 35:60-62.

Bassolos, M. C. and J. N. Moore. 1981. 'Ebano' thornless blackberry. *HortScience* 16:686-687.

Braun, J. W. and J. K. L. Garth. 1984. Intracane yield compensation in the red raspberry. *J. Amer. Soc. Hort. Sci.* 109:526-530.

Brierley, J. W. 1934. Studies of the response of Latham raspberry to pruning treatment. Minn. Agr. Exp. Sta. Tech. Bul. 100.

Bristow, P. R., H. A. Daubeny, T. M. Sjulin, H. S. Pepin, R. Nesby and G. E. Windom. 1988. Evaluation of *Rubus* germplasm for reaction to root rot caused by *Phytophthora erythroseptica*. *J. Amer. Soc. Hort. Sci.* 113:588-591.

Brooks, R. M. and H. P. Olmo. 1944. Register of new fruit and nut varieties. *Proc. Amer. Soc. Hort. Sci.* 45:467-490.

Brun, C. A. 1992. Growing small fruits for the home garden. Wash. Coop. Ext. Serv. EB 1640.

Bullock, R. M. 1963. Spacing and time of training blackberries. *Oreg. Hort. Soc.* 55:59-60.

Burr, T. J., C. L. Reid, B. H. Katz, M. E. Tagliati, C. Bazzi and D. I. Breth. 1993. Failure of *Agrobacterium radiobacter* strain K-84 to control crown gall on raspberry. *HortScience* 28: 1017-1019.

Caldwell, J. D. 1984. Blackberry propagation. *HortScience* 19: 193-195.

Christensen, J. R. 1947. Root studies XI. Raspberry root systems. *J. Pomol.* 23:218-226.

Clark, J. R. 1992. Blackberry production and cultivars in North America east of the Rocky Mountains. *Fruit Var. J.* 46:217-222.

Colby, A. S. 1936. Preliminary report on raspberry root systems. *Proc. Amer. Soc. Hort. Sci.* 34:372-376.

Converse, R. H. (Ed.). 1987. *Virus Diseases of Small Fruits.* U.S.D.A. Agr. HB 631.

Converse, R. H. 1991. Diseases caused by viruses and virus-like agents, In *Compendium of Raspberry and Blackberry Diseases and Insects,* M. A. Ellis, R. H. Converse, R. N. Williams and B. Williamson (Eds.). APS Press: St. Paul, MN.

Crandall, P. C. 1980. Twenty years of red raspberry research in Southwestern Washington. *Acta Hort.* 112:53-57.

Crandall, P. C. and J. D. Chamberlain. 1972. Effects of water stress, cane size and growth regulators on floral primordia development in red raspberries. *J. Amer. Soc. Hort. Sci.* 97:418-419.

Crandall, P. C. and H. A. Daubeny. 1990. Raspberry management. In *Small Fruit Crop Management,* G. J. Galletta and D. G. Himelrick (Eds.). Prentice-Hall: Englewood, NJ.

Crandall, P. C., J. D. Chamberlain and K. A. Biderbost. 1974. Cane characteristics associated with berry number of red raspberry. *J Amer. Soc. Hort. Sci.* 99:370-372

Crandall, P. C., J. D. Chamberlain and J. K. L. Garth. 1980. The effects of chemical primocane suppression on growth, yield and chemical composition of red raspberries. *J. Amer. Soc. Hort. Sci.* 105:194-196.

Crandall, P. C., D. F. Allmendinger, J. D. Chamberlain and K. A. Biderbost. 1974. Influence of cane number and diameter, irrigation and carbohydrate reserves on the fruit number of red raspberries. *J. Amer. Soc. Hort. Sci.* 99:524-526.

Crandall, P. C., M. C. Jensen, J. E. Middleton and J. D. Chamberlain. 1969. Scheduling the irrigation of red raspberries from evaporation data. Wash. Agr. Exp. Sta. Cir. 497.

Dale, A. 1989. Productivity of red raspberries. *Hort. Rev.* 11:185-228.

Dale, A. 1992. Raspberry cultivars in eastern Canada. *Fruit Var. J.* 46:222-225.

Dale, A. and H. A. Daubeny. 1985. Genotype-environment interaction involving British and Pacific Northwest raspberry cultivars. *HortScience* 20:68-69.

Dale, A. and H. A. Daubeny. 1987. Flower-bud initiation in red raspberry (*Rubus idaeus* L.) in two environments. *Crop Res.* 27:61-66.

Dana, M. and B. Goulart. 1989. Bramble biology. In *Bramble Production Guide*, M. Pritts and D. Handley (Eds.). N. E. Reg. Agr. Eng. Serv. NRAES-35.

Darrow, G. M. 1931. A productive thornless sport of the Evergreen blackberry. *J. Hered.* 22:405-406.

Darrow, G. M. 1937. Blackberry and raspberry improvement. In *Better Plants and Animals*. U.S.D.A. Yearbook of Agriculture.

Daubeny, H. A. 1986. The British Columbia raspberry breeding program since 1980. *Acta Hort.* 183:47-58.

Daubeny, H. A. 1991. Raspberry. In Register of new fruit and nut varieties. List 35. J. N. Cummins (Ed.). *HortScience* 26:978-980.

Daubeny, H. A. and C. Fear. 1992. Primocane fruiting raspberries in the Pacific Northwest and California. *Fruit Var. J.* 46:197-199.

Daubeny, H. A. and D. Stary. 1982. Identification of resistance to *Amphorophora agathonica* in the native North American red raspberry. *J. Amer. Soc. Hort. Sci.* 107:593-597.

Daubeny, H. A., P. C. Crandall and G. W. Eaton. 1967. Crumbliness in the red raspberry with special reference to the 'Sumner' variety. *Proc. Amer. Soc. Hort. Sci.* 91:224-230.

Daubeny, H. A., F. J. Lawrence and G. R. McGregor. 1989. 'Willamette' red raspberry. *Fruit Var. J.* 43:46-48.

Daubeny, H. A., R. Stace-Smith and J. A. Freeman. 1978. The occurrence and some effects of raspberry bushy dwarf virus in red raspberry. *J. Amer. Soc. Hort. Sci.* 103:519-522.

Daum, D. R. and T. F. Reed. 1983. Boom sprayers. Penn. Coop. Ext. Serv. NRAES-19.

Debney, H. G., K. J. Blacker, B.J. Redding and J. B. Watkins. 1980. *Handling and Storage Practices for Fresh Fruit and Vegetables. Product Manual.* Australian United Fresh Fruit and Vegetable Assoc.

Donnelly, D. J. and H. A. Daubeny. 1986. Tissue culture of *Rubus* species. *Acta Hort.* 183:305-314.

Doughty, C. C., P. C. Crandall and C. H. Shanks, Jr. 1972. Cold injury to red raspberries and the effect of premature defoliation and mite damage. *J. Amer. Soc. Hort. Sci.* 97:670-673.

Eaton, G. W., P. A. Bowen and P. A. Jolliffe. 1986. Two-dimensional partitioning of yield variation. *HortScience* 21:1052-1053.

Ellis, M. A., R. H. Converse, R.N. Williams and B. Williamson (Eds.). 1991. *Compendium of Raspberry Diseases and Insects.* APS Press: St. Paul, MN.

Fear. C. D. and L. B. Hertz. 1982. Effects of NAA on cane growth and yield of 'Boyne' red raspberries. 1984. *HortScience* 17:1770.

Focke, W. O. 1910-1914. Species Ruborum Bibliotheca. *Botanica* 72:1-233; 83:1-274. Schweizerbartsche Verlagsbuchhandlung, Stuttgart.

Frehman, S. 1989. Living mulch ground covers for weed control between raspberry rows. *Acta Hort.* 262:349-352.

Galletta, G. J. and D. G. Himelrick (Eds.) 1990. *Small Fruit Crop Management.* Prentice-Hall, Inc.: Englewood Cliffs, NJ.

Galletta, G. J. and C. Viollette. 1989. The brambles. In *Bramble Production Guide*, M. Pritts and D. Handley (Eds.). N.E. Reg. Argo. Eng. Serv. NRAES-35.

Gerber, H. S. 1984. *Major Insect and Mite Pests of Berry Crops.* B.C. Min. Agr. and Foods.

Gerhardt, S. E., R. Cutrufelli and R. H. Matthews. 1982. *Composition of Foods, Fruits and Fruit Juices–Raw, Processed, Prepared.* U.S.D.A. Human Nutrition Services. Agr. HB 8-9.

Green, A. 1971. Soft fruits. In *The Biochemistry of Fruits and their Products*, A. C. Hulme (Ed.). Academic Press: London and New York.

Hall, H. K. and L. R. Brewer. 1989. Breeding *Rubus* cultivars for warm temperature climates. *Acta Hort.* 262:65-74.

Harper, P. C. 1978. Tissue culture propagation of blackberry and tayberry. *Hort. Res.* 18:141-143.

Harris, R. W. and R. H. Coppock. 1978. Saving water in landscape irrigation. Univ. Calif. Agr. Sci. Leaflet 2976.

Hedrick, U. P. 1925. *The Small Fruits of New York.* J. B. Lyon: Albany, NY.

Hill, R. G., Jr. 1958. Fruit development of the red raspberry and its relation to nitrogen treatment. Ohio Agr. Exp. Sta. Res. Bul. 803.

Hoover, E., J. Luby and D. Bedford. 1986. Yield components of primocane fruiting red raspberries. *Acta Hort.* 183:163-166.

Hudson, J. P. 1959. Effects of environment on *Rubus idaeus* L. I. Morphology and development of the raspberry plant. *J. Hort. Sci.* 34:163-169.

Jennings, D. L. 1981. A hundred years of Loganberries. *Fruit Var. J.* 35:34-37.

Jennings, D. L. 1988. *Raspberries and Blackberries: Their Breeding, Diseases and Growth.* Academic Press: London and New York.

Jennings, D. L. and A. Dale. 1982. Variation in the growth habit of red raspberries with particular reference to cane height and node production. *J. Hort. Sci.* 57:197-204.

Jennings, D. L. and B. M. M. Tulloch. 1965. Studies of factors which promote germination of raspberry seeds. *J. Exp. Hort.* 16:329-340.

Jennings, D. L., H. A. Daubeny and J. N. Moore. 1991. Blackberries and raspberries. In *Genetic Resources of Temperate Fruit and Nuts.* J. N. Moore and J. R. Ballington (Eds.) Int. Soc. Hort. Sci., Wageningen.

Johnson, F., P. C. Crandall and J. R. Fisher. 1972. Soil fumigation and its effect on raspberry root rot. *Pl. Dis. Reptr.* 56:467-470.

Keep, E. 1984. Breeding *Rubus* and *Ribes* crops at East Malling. *Sci. Hort.* 35:54-71.

Knight, A. L., R. LaLone, G. C. Fisher and L. B. Coop. 1988. Managing leafrollers on caneberries in Oregon. Oreg. Coop. Ext. Serv. Ext. Cir. 1263.

Lawrence, F. J. 1980. The current status of red raspberry cultivars in the United States and Canada. *Fruit Var. J.* 34: 84-89.

Lipe, J. A. 1986. Keys to profitable blackberry production in Texas. Texas Agr. Exp. Sta. Ser. B-1560.

Locklin, H. A. 1932. Effects of different methods of pruning raspberries on earliness, weight of fruit and yield. *Proc. W. Wash. Hort. Assoc.* 27:185-189.

McGregor, G. R. and K. H. Kroon. 1984. Silvan blackberry. *HortScience* 19:732-733.

McPheeters, K. R., R. M. Skirvin and H. K. Hall. 1988. Brambles (*Rubus* spp.) tissue culture. In *Biotechnology in Agriculture and Forestry*, Y.P.S. Bajaj (Ed.). Vol. 6 Crops II., Springer-Verlag.

Moore, J. N. 1980. Blackberry production and cultivar situation in North America. *Fruit Var. J.* 34:36-41.

Moore, J. N. 1984. Blackberry breeding. *HortScience*. 19:183-197.

Moore, J. N. and J. D. Caldwell. 1983. *Rubus* (Tourn.) L. In *Genetic Resources of Temperate Fruit and Nuts*. J. N. Moore and J. P. Ballington (Eds.) Int. Soc. Hort. Sci., Wageningen: Netherlands.

Moore, J. N. and R. M. Skirvin. 1990. Blackberry management. In *Small Fruit Crop Management*, G. J. Galletta and D. G. Himelrick (Eds.). Prentice-Hall: Englewood, NJ.

Moore, J. N., G. C. Paulis, G. R. Brown and C. R. Lindergan. 1978. Establishing blackberry plantings with root cuttings. *Ark. Farm Res.* 27:4.

Moore, P.P. 1992. Floricane fruiting red raspberry cultivars in the Pacific Northwest. *Fruit Var. J.* 46:200-202.

Nakasu, B. H., M. C. Bassolos and A. J. Feliciano. 1981. Temperate fruit breeding in Brazil. *Fruit Var. J.* 35:114-122.

Nelson, G. S., R. H. Benedict, A. A. Kattan and G. A. Albritton. 1965. Design and development of a blackberry harvester. *Amer. Soc. Agr. Eng.* Paper 65-69.

Nonnecke, G. R. and J. J. Luby. 1992. Raspberry cultivars and production in the midwest. *Fruit Var. J.* 46:207-212.

Perkins-Veazie, P. and G. R. Nonnecke. 1992. Physiological changes during ripening of raspberry fruit. *HortScience* 27:331-333.

Pritts, M. 1989. Pruning and trellising brambles. In *Bramble Production Guide*. M. Pritts and D. Handley (Eds.). N.E. Reg. Agr. Eng. Serv. NRAES-35.

Pritts, M. and D. Handley. 1989. *Bramble Production Guide*. N.E. Reg. Agr. Eng. Serv. NRAES-35.

Reeve, R. M. 1954. Fruit histogenesis in *Rubus strigosus. Amer. J. Bot.* 41:152-160; 173-181.

Robbins, J. A. and P. P. Moore. 1991. Fruit morphology and fruit strength in a seedling population of red raspberry. *HortScience* 26:294-295.

Robbins, J. A. and T. M. Sjulin. 1988. Scanning electron microscope analysis of drupelet morphology of red raspberry and related *Rubus* genotypes. *J. Amer. Hort. Sci.* 113:474-480.

Robertson, M. 1957. Further investigations of flower bud development in the genus *Rubus*. *J. Hort. Sci.* 32:265-273.

Rodriguez, J. A. and E. G. Avitia. 1989. Advances in breeding low-chill red raspberries in central Mexico. *Acta Hort.* 262: 127-132.

Ross, D. S. 1987. Pesticide sprayers for small farms. Maryland Coop. Ext. Serv. Bul. 317.

Scheer, W. P. A. and R. Garren. 1981. Commercial red raspberry production. Wash. Coop Ext. Serv. PNW Bul. 176.

Scheer, W. P. A., C. A. Brun, C. B. MacConnell and A. L. Antonelli. 1993. 1993 pest control guide for commercial small fruits. Wash. Coop. Ext. Serv. EB 1491.

Schwartze, C. D. and A. S. Myhre. 1952. Some experiments and observations on mechanical harvesting of red raspberries. *Proc. W. Wash. Hort. Assoc.* 42:21-24.

Shanks, C. H., Jr. 1969. Pollination of raspberries by honey bees. *J. Apic. Res.* 8:19-21.

Shanks. C. H., Jr., J. E. George and P. C. Crandall. 1972. A hooded spray boom for caneberries, bush or vine fruit. Wash. Coop. Ext. Serv. EB 638.

Sheets, W. A., T. L. Nelson and A. G. Nelson. 1975. Alternate-year production of 'Thornless Evergreen' blackberries. Technical and economic feasibility. Oregon Agr. Exp. Sta. Res. Bul. 620.

Shoemaker, J. S. 1978. *Small Fruit Culture*. AVI: Westport, CN.

Sjulin, T. M. and J. A. Robbins. 1987. Effects of maturity, harvest date and storage time on postharvest quality of red raspberry fruit. *J. Amer. Soc. Hort. Sci.* 122:481-487.

Snir, I. 1988. Red raspberry (*Rubus idaeus*). In *Biotechnology in Agriculture and Forestry,* Y. P. S. Bajaj (Ed.). Vol. 6. Crops II. Springer-Verlag.

Spayd, S. E., J. R. Morris, W. E. Ballinger and D. G. Himelrick. 1990. In *Small Fruit Crops Management,* G. J. Galletta and D. G. Himelrick (Eds.). Prentice-Hall: Englewood, NJ.

Strik, B. C. 1992. Blackberry cultivars and production trends in the Pacific Northwest. *Fruit Var. J.* 46:202-206.

Swartz, H. J., S. E. Gray, L. W. Douglass, E. Durner, C. S. Walsh and G. J. Galletta. 1984. The effect of a divided canopy trellis design on thornless blackberry. *HortScience* 26:533-535.

Swartz, H. J., S. K. Naess, J. Fiola, H. Stiles, B. Smith, M. Pritts, J. C. Sanford and K. Maloney. 1992. Raspberry genotypes for the East Coast. *Fruit Var. J.* 46:212-216.

Takeda, F. 1987. Some factors associated with fruit maturity range in cultivars of the semi-erect tetraploid thornless blackberries. *HortScience* 22:405-408.

Takeda, F. and M. Wisniewski. 1989. Organogenesis and patterns of floral bud development in two eastern thornless blackberry cultivars. *J. Amer. Soc. Hort. Sci.* 114:528-531.

Torre, L. C. and B. H. Barritt. 1979. Red raspberry establishment from root cuttings. *J. Amer. Soc. Hort. Sci.* 104:28-31.

Turner, D. and K. Muir. 1985. *The Handbook of Soft Fruit Growing.* Croom Helm Ltd.: London.

Vasilakakis, M. D., B. H. McCown and M. N. Dana. 1979. Hormonal changes associated with growth and development of red raspberries. *Physiol. Plantarum* 45:17-22.

Vasilakakis, M. D., B. E. Struckmeyer and M. N. Dana. 1979. Low temperature and development of red raspberries. *HortScience* 104:61-62.

Waister, P. D. 1970. Effects of shelter from wind on the growth and yield of raspberries. *J. Hort. Sci.* 45:435-445.

Waister, P. D. and B. H. Barritt. 1980. Compensation in fruit numbers following loss of lateral buds in the red raspberry. *Hort. Res.* 20:25-31.

Waister, P. D., N. R. Cormack and W. A. Sheets. 1977. Competition between fruit and vegetative phases in red raspberry. *J. Hort. Sci.* 52:75-85.

Waister, P. D., C. J. Wright and M. R. Cormack. 1980. Potential yield in red raspberry as influenced by interaction between genotype and cultural methods. *Acta Hort.* 112:273-283.

Waldo, G. F. 1933. Fruit bud formation in brambles. *Proc. Amer. Soc. Hort. Sci.* 30:263-267.

Walsh, C. S., J. Popenoe and J. T. Solomos. 1983. Thornless blackberry is a climacteric fruit. *HortScience* 11:515-517.

Williams, I. H. 1959a. Effects of environment on *Rubus idaeus* L. III. Growth and dormancy of young shoots. *J. Hort. Sci.* 34:210-218.

Williams, I. H. 1959b. Effects of environment on *Rubus idaeus* L. IV. Flower initiation and development of the inflorescence. *J. Hort. Sci.* 34:219-228.

Williams, I. H. 1960. Effects of environment on *Rubus idaeus* L. V. Dormancy and flowering of the mature shoot. *J. Hort. Sci.* 35:214-220.

Index

Milton Keynes UK
Ingram Content Group UK Ltd.
UKHW050259161024
449569UK00042B/1811